BARRON'S

Painless
Earth
Science

SECOND EDITION

Edward J. Denecke Jr.

Published by Kaplan, Inc., d/b/a Barron's Educational Series
750 Third Avenue
New York, NY 10017
www.barronseduc.com

ISBN: 978-1-5062-7322-8

10 9 8 7 6 5 4 3 2

Kaplan, Inc., d/b/a Barron's Educational Series print books are available at special quantity discounts to use for sales promotions, employee premiums, or educational purposes. For more information or to purchase books, please call the Simon & Schuster special sales department at 866-506-1949.

Contents

How To Use This Book

Painless Earth Science? *Impossible*, you think. Not really. Earth Science is easy… or at least it can be with the help of this book!

Earth science encompasses the study of Earth's land, water, air, and life, and how they interact. As you read this book, you will probably be surprised at how much of Earth science is already familiar to you. That's because Earth science is all around you, and it affects every aspect of your life. Whether you are learning Earth science for the first time, or you are trying to remember what you've learned but have forgotten, this book is for you. Don't be afraid. Dive in—it's painless!

Painless Icons and Features

This book is designed with several unique features to help make learning Earth science easy.

 **PAINLESS TIP**

You will see Painless Tips throughout the book. These include helpful tips, hints, and strategies on the surrounding topics.

 **CAUTION—Major Mistake Territory!**

Caution boxes will help you avoid common pitfalls or mistakes. Be sure to read them carefully.

 BRAIN TICKLERS

There are Brain Ticklers throughout each chapter in the book. These quizzes are designed to make sure you understand what you've just learned and to test your progress as you move forward in the chapter. Complete all the Brain Ticklers and check your answers. If you get any wrong, make sure to go back and review the topics associated with the questions you missed.

ILLUSTRATIONS

Painless Earth Science is full of illustrations to help you better understand Earth science topics. You'll find tables, graphs, charts, and instructive science illustrations to help you along the way.

SIDEBARS

These shaded boxes contain extra information that relates to the surrounding topics. Sidebars can include more advanced topics, detailed examples, and more to help keep Earth science interesting and painless.

Chapter 1

Earth's Structure

Earth's Spheres

Spheres, spheres, spheres! The first thing you need to know about Earth is that it is made up of lots of layers and that each of those layers is shaped like a sphere. When Earth first formed it was molten, and gravity pulling toward its center caused it to form a sphere. Like a mixture of oil and water, the substances that made up Earth then separated into layers due to density differences. Gravity caused denser substances like rock to sink inward toward the center and less dense substances like gases to float outward toward the surface. Since Earth is shaped like a sphere, the layers that formed are also spheres. You can think of Earth as a whole bunch of spheres, one inside the other. Earth's major layers, or spheres, include the lithosphere, hydrosphere, atmosphere, and biosphere. Let's take a closer look at each.

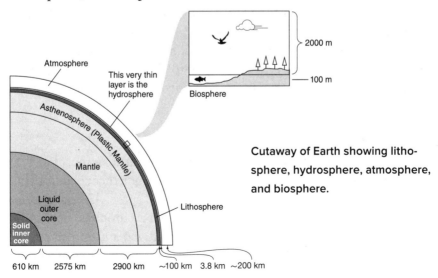

Cutaway of Earth showing lithosphere, hydrosphere, atmosphere, and biosphere.

Lithosphere

We often think of the entire Earth beneath our feet as solid. But think about a volcanic eruption. Clearly, liquid rock is coming up from below Earth's surface. So, at least part of what lies beneath the surface is not solid. Scientific investigation has revealed that the rock inside Earth is hot enough to flow slowly like melted butter and has also separated into layers according to density. The **lithosphere** is Earth's cold, hard, solid outer layer of soil and rock extending from the surface to a depth of about 100 kilometers. Beneath the lithosphere are more layers of hot rock reaching all the way down Earth's center more than 6,300 kilometers beneath the surface. All of Earth's mountains, valleys, plains, plateaus, and other surface features are part of the lithosphere.

Hydrosphere

When you dive into the water at the beach, you are diving into the hydrosphere. The **hydrosphere** is a thin layer of liquid water that rests upon the lithosphere. More than 70 percent of Earth's surface is covered by water. The trillions of gallons of water that make up the hydrosphere cover all of the low spots in Earth's lithosphere to an average depth of about 3.8 kilometers (2.4 miles). This is very thin compared to Earth's diameter. If you dipped a basketball in water, the water wetting its surface would be deeper in places than the hydrosphere is on Earth.

The hydrosphere plays a key role in many geologic processes. Moving water carries loose rock from place to place and shapes Earth's surface. The oceans act as heat absorbers, preventing drastic temperature changes. Water is also essential to all living things, not only as drinking water, but as the main substance in the cells of all living things.

Atmosphere

Every time you take a breath, you are breathing Earth's atmosphere. The **atmosphere** is a thin layer of air that surrounds the whole Earth and extends out several hundred kilometers into space. **Air** is a mixture made up mostly of gases, but it also contains water droplets, ice, dust, and other particles. Air is about 78 percent nitrogen and 21 percent oxygen. The remaining 1 percent is mostly argon with traces of carbon dioxide and other gases. The atmosphere also contains water vapor, but the amount varies from 0 percent over deserts to as much as 4 percent over tropical jungles. All of Earth's weather, from puffy little clouds to massive hurricanes, occurs in the atmosphere.

Biosphere

You and all of your friends are part of the biosphere. So is the grass in your lawn, the trees in the park, your pet dog, and the fleas and ticks on your dog. Even the bacteria and viruses that make you sick are part of the biosphere. The **biosphere** consists of all life on Earth. It may seem odd to think of life as a sphere, but think of what Earth would look like if you stripped away everything that is nonliving. Earth is surrounded by a thin layer of life that exists on and in its land, throughout its water, and in the lower parts of its air. The presence of a biosphere and its interaction with the other spheres makes Earth a unique planet.

PAINLESS TIP

litho means rock   *hydro* means water

atmo means air   *bio* means life

So, biosphere means "life sphere," lithosphere means "rock sphere," hydrosphere means "water sphere," and atmosphere means "air sphere."

When we say something is a **system**, we mean it has parts that are interdependent and interact within the system. A cell phone is a good example of a system. It has many parts that interact and depend on one another—buttons, screen, microphone, and speaker—to name just a few. You can think of Earth as a system of interacting spheres, one inside the other. Fish (biosphere) swim through the oceans (hydrosphere). During a storm (atmosphere) rain falls to the ground (lithosphere) and may run off into a stream (hydrosphere) where a deer (biosphere) is drinking.

Whenever the spheres of Earth system interact, changes occur. For example, when rain falls to the ground during a storm and runs off, it erodes the land and changes the land's shape. Changes always involve a transfer of energy from one part of the system to another. Most of the energy in Earth system can be traced back to the Sun. For instance, the water that fell as rain in the storm got into the atmosphere when the energy in sunlight caused it to evaporate.

YOU ARE PART OF THE BIOSPHERE

You and all humans are living organisms. Therefore, you are part of the biosphere. Humans have had far-reaching effects on all of Earth's spheres.

Humans sometimes damage or destroy natural habitats causing the extinction of other species in the biosphere.

Humans have changed the atmosphere by burning fossil fuels such as coal, oil, and natural gas, releasing vast amounts of carbon dioxide into the atmosphere. The increased carbon dioxide in the atmosphere traps heat and has warmed the atmosphere. This, in turn, has caused warming of the oceans and melting of ice in the hydrosphere.

Humans have even changed the lithosphere by changing its land structure. Humans have excavated, paved, cleared, and reshaped the land, often resulting in increased erosion. They have polluted the ground with human-made chemicals and nuclear radiation.

However, human activities can have positive impacts if the activities and technologies we use are engineered differently with conservation in mind.

The spheres of Earth system are also interdependent. That means they affect one another. For example, the location and depth of a lake (hydrosphere) is affected by the shape of Earth's surface (lithosphere). In turn, the shape of Earth's surface is affected by erosion by streams and waves (hydrosphere). The lithosphere affects the hydrosphere and the hydrosphere affects the lithosphere—the lithosphere and hydrosphere are interdependent.

PAINLESS TIP

When trying to predict how substances will move in Earth system, ask yourself, "Which substance is more dense than its surroundings?" "Which is less dense?" If it is more dense than its surroundings, it will tend to move toward Earth's center, or sink. If it is less dense, it will tend to move away from Earth's center, or float.

BRAIN TICKLERS   Set # 1

Place an L, H, A, or B next to each item to indicate if it is part of the lithosphere, hydrosphere, atmosphere, or biosphere, respectively.

1. Boulder   _____

2. Tornado   _____

3. River   _____

4. Bird   _____

5. Water in a well   _____

6. Gold mine   _____

7. Ocean wave   _____

8. Whale   _____

9. Which set of Earth components is arranged in order from solid to liquid to gas?

   a. hydrosphere, atmosphere, lithosphere

   b. hydrosphere, lithosphere, atmosphere

   c. lithosphere, atmosphere, hydrosphere

   d. lithosphere, hydrosphere, atmosphere

10. Which diagram best shows Earth with the hydrosphere drawn to scale?

a. Ocean → (Earth)   b. Ocean → (Earth)   c. Ocean → (Earth)   d. Ocean → (Earth)

11. Describe an event that involves an interaction between two spheres of Earth system. Among three spheres. Among four spheres.

(Answers are on page 28.)

Earth Models

Making a model of Earth is a good way to begin studying it. A model is anything that represents the properties of an object or a system. A model is both alike and different from a real thing, but it can be used to learn something about the real thing. A model may be an object, such as a statue to represent a human or a globe to represent Earth. Or a model may be a drawing, a photograph, a chart, or a table. A model can even be a mathematical equation in which symbols such as words or numbers are used to represent objects or parts of systems, and the ways in which they are related to one another. Models are often used to think about things that are too big or too small, or that happen too quickly or too slowly, to be observed or changed directly. For example, models are often used to represent Earth because it is too big to be observed directly by a person on its surface. Models may also be used to study things that could be dangerous. That is why scientists use models of a building to study how they are affected by earthquakes. Observing how a model responds after it is changed may suggest how the real thing would respond if the same thing were done to it.

PAINLESS TIP

Whenever you think of models, think of the letters **MVP**—for mathematical, visual, and physical, respectively.

Mathematical models are like equations that represent relationships between parts of a system.

Visual models are things like photographs, diagrams, or charts.

Physical models are things like action figures, toy cars, and dollhouses.

Scale

Most models are made to **scale**; that is, the parts of the model are made in the same proportions as the parts in the original. A statue of a man would look quite odd if its legs were one-tenth the size of a real leg, but its arms were one-half the size of a real arm. A model's scale is the ratio of the size of the model part to the original part. A map drawn to a scale of 1:62,500 means that one unit of distance on the map is equal to 62,500 units of distance on the ground. Both numbers in the ratio have the same units, but those units could be anything. For example, 1 inch on the map would equal 62,500 inches on the ground (5,208 feet or roughly a mile). It also means that 1 centimeter on the map equals 62,500 centimeters on the ground. In order to create a scale model that represents Earth correctly we need to know its shape and size.

PAINLESS TIP

Think of a doll and a person. If made to scale, the parts of the doll's body will be the same size in relation to one another as a human's body. The parts of the doll should also have the same shape as a human's parts.

Earth's shape and size

Earth's shape is almost, but not quite, a perfect sphere. A perfect sphere would have exactly the same diameter when measured in any direction. Earth's actual measurements differ slightly. Earth's diameter through the poles is 12,714 kilometers. Earth's diameter through the equator is a little larger: 12,756 kilometers. Thus, Earth's spherical shape "bulges" very slightly at the equator and is very slightly "flattened" at the poles. This shape is called an **oblate** (flattened) **spheroid**. However, the shape of Earth is so close to being a perfect sphere that your eye would not be able to detect its oblateness. Let's suppose that we made a scale model of Earth—a globe. If we used a scale of 1 centimeter = 1,000 kilometers, the globe would have a polar diameter of 12.714 centimeters and an equatorial diameter of 12.756

centimeters. This is a little bigger than a softball. The difference in diameters would be 0.042 centimeter, or less than a half a millimeter. You would need a micrometer to measure the difference in diameters. Any cross-section of Earth looks like a perfect circle.

Earth's oblateness is the result of forces produced by Earth's rotation on its axis. Just as a loose skirt will swirl outward if the person wearing it spins around, Earth "swirls" outward when it rotates. However, since Earth is much stiffer than a skirt, the distance it moves outward is much smaller.

It would seem then that a globe would be the best model of Earth. But there are problems with a globe. You cannot see an entire globe at once. One half is always facing away from you. If you make a globe at a scale that would show the details of a small region (like a city), the globe would be too large to manage. One solution is to use a different kind of model—a map.

PAINLESS TIP

When thinking of Earth, think of a perfectly round ball. Earth's shape is so close to being a perfect sphere that its "out of roundness" can only be detected with sensitive instruments.

BRAIN TICKLERS   Set # 2

1. List three examples of different kinds of models.

_____

_____

_____

2. Give two reasons why a scientist might use a model to study something instead of just studying the real thing.

_____

_____

3. Earth's shape most closely resembles which of the following objects?

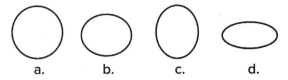

a.   b.   c.   d.

4. At sea level, which would be farthest from Earth's center?

    a. The north pole   **c.** 45° north latitude

    b. 23 1/2° south latitude   **d.** The equator

5. A student builds a model car from a kit. The kit box says that the scale of the model is 1:25. The finished model car is 15 centimeters long. How long is the actual car?

    a. 25 cm   **b.** 40 cm   **c.** 375 cm   **d.** 750 cm

6. Explain why a globe is not always the best model to use for Earth, even though Earth is a sphere.

(Answers are on page 28.)

Mapping Earth's Surface

A map is a type of visual model, usually drawn on a flat surface, which represents the features of an area. A map is meant to communicate a sense of place, of where one point is in relation to another. A map can be anything from a quick sketch showing a friend how to get to the park from school to an elaborate scale model of Earth complete with mountain ranges and ocean basins. Although flat maps differ from the three-dimensional Earth, they are a useful tool for learning about the things they represent. The nature of a map depends on the purpose for which it was created.

Earth scientists use many different types of maps. Some maps show Earth's surface features or the type of bedrock found at the surface. Others show the depth of Earth's waters, or weather systems in the atmosphere. There are even maps that show where the stars in the universe appear at night.

Locating positions

In order to construct a map of Earth's surface, you need to have
a way of determining where things are located in relation to one
another. Imagine you are going to visit a museum in New York City.
To find your way around the city you would look at a map. The map
would show that streets run east to west and avenues run north to
south. The streets and avenues form a grid that makes finding places
simple. The New York City street-avenue grid is a type of coordinate
system. A **coordinate system** is a way of locating points by labeling
them with numbers called coordinates. **Coordinates** are numbers
measured with respect to a system of lines or some other fixed reference. When you say that you are at the corner of 5th Avenue and
42nd Street, you are giving your location using coordinates. Now
think about how you would locate places on Earth where there is no
grid—like the middle of the ocean!

Latitude-Longitude

Scientists have solved this problem by using *imaginary* grid lines.
In order to describe the position of any point on Earth's spherical
surface, they set up a coordinate system that uses two coordinates
known as latitude and longitude. The latitude-longitude system is
made up of two sets of imaginary lines that cross each other at right
angles.

Find the equator in the figure on page 11. The **equator** is a line circling Earth halfway between the north and south poles. Notice the
lines drawn north and south of the equator. These are latitude lines.
Since Earth is a sphere, these latitude lines actually form circles.
The circles formed by latitude lines are called **parallels**, because if
you drew a series of latitude lines, the circles formed would all be
parallel to one another. Notice, though, that the farther the latitude
is from the equator, the smaller the circle. Latitude lines are labeled
in degrees by their angular distance north or south of the equator as
measured from the center of Earth. The equator is the reference line,
or starting point; therefore, latitude is 0° at the equator and 90° at
the poles.

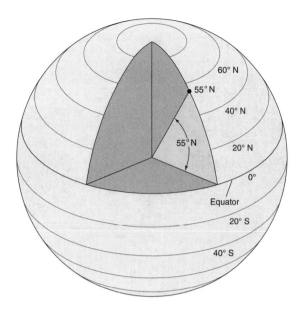

Cutaway drawing showing the latitude angle at Earth's center for a point at
55°N latitude.

In the figure on the following page, find the prime meridian. The
prime meridian is a line drawn from pole to pole passing through
Greenwich, England. Notice the lines drawn east and west of the
prime meridian. These are longitude lines. Again, since Earth is a
sphere, when longitude lines are extended they form circles. Any circle that passes through both of Earth's poles is called a **meridian**.

Notice that like latitude lines, longitude lines are also labeled in
degrees. Longitude lines are labeled by their angular distance east or
west of the prime meridian as measured from the center of Earth.

Unlike the equator, which is the *only* line halfway between the poles,
the reference line for longitude, or the prime meridian, could be
any of the meridians because they're all the same. In 1884, an inter-
national conference agreed that the prime meridian would be the
meridian of longitude that runs through the Royal Observatory in
Greenwich, England. The prime meridian is labeled 0° longitude.
Longitude is measured in degrees east or west of the prime meridian
up to 180°.

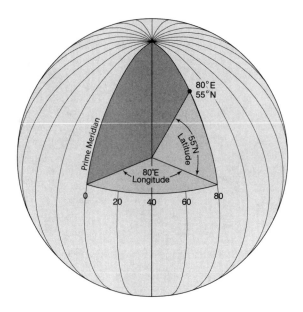

Cutaway drawing showing the longitude angle at Earth's center.

PAINLESS TIP

On most maps, north is toward the top, south toward the bottom, east to the right, and west to the left. North latitudes increase toward the top and south latitudes increase toward the bottom. East longitudes increase to the right and west longitudes increase to the left.

Any meridian will cross the equator and all other parallels at right angles. So, once reference lines are chosen, meridians and parallels form a grid of lines that intersect at right angles on the surface of Earth even though it is a sphere—a pretty neat coordinate system! This is called the latitude-longitude coordinate system. Every point on Earth's surface can be described by the latitude and longitude lines that intersect at that point. The coordinates of any point is given as the number of degrees north or south of the equator of the latitude line and the number of degrees east or west of the prime meridian of the longitude line that intersect at that point. For example, the coordinates of Mount Everest are 28°N, 87°E.

Latitude-longitude grid on spherical Earth showing coordinates for Mt. Everest.

PAINLESS TIP

Latitude-longitude coordinates are always given with the latitude first and then the longitude.

Each coordinate consists of a number of degrees followed by the capital letter N or S for north or south latitude, respectively; or the capital letter E or W for east or west longitude, respectively.

Map projections

If you try to make a map of a large region of Earth's surface, you run into problems. Earth's surface is curved and a map is flat. Imagine trying to get a basketball to lie flat on a table. To make the basketball lie flat, you have to stretch it. One way to solve this problem is to make a map projection. In a map projection, features of Earth's curved surface are projected onto a flat surface like shadows on a wall. Notice the different types of projections shown on page 14. Each has its advantages, but each also stretches or distorts Earth's features in some way.

Transverse Mercator

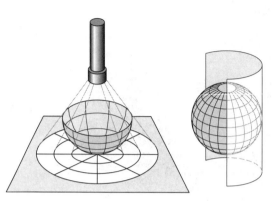

Map projections.

BRAIN TICKLERS   Set # 3

1. Which of the following best shows Earth's latitude-longitude coordinate system?

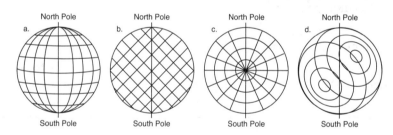

2. List one way in which parallels and meridians are the same and two ways in which they are different.

Same _____

Different _____; _____

3. In the map below, which location is in the shaded area of basalt bedrock?

a. 48°N, 115°W

b. 46°N, 118°W

c. 44°N, 122°W

d. 40°N, 120°W

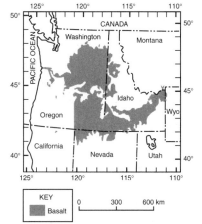

4. The diagram below shows part of Earth's latitude-longitude system. What is the latitude and longitude of point *L*?

a. 5°N, 30°E

b. 5°W, 30°S

c. 5°S, 30°W

d. 5°E, 30°N

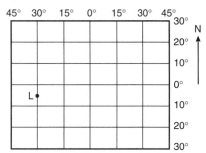

5. A is a way of representing a round globe on a flat map.

a. coordinate system

b. map projection

c. prime meridian

d. parallel

(Answers are on page 28.)

Field Maps

Earth's surface is not the only part of Earth system that scientists map. They also create maps that help them visualize properties of the atmosphere, hydrosphere, and lithosphere such as temperature, pressure, and motion. But how do you map something like a property? The solution is to create a **field map**.

A **field** is a region of space that has a measurable property at every point. Some examples of field properties are temperature, pressure, magnetism, gravity, and elevation. Field maps can be used to represent any property that varies in a region of space. For example, you could use a thermometer to measure the temperature at every single point in your classroom. You could create a map by measuring the temperature at lots of different locations and then plotting the temperatures on a map of your classroom. The result would be a temperature field map. A field map simply shows how much of something (temperature in this case) is found at many locations.

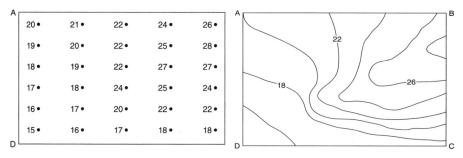

Temperature field map of a classroom. Left: plotted points; right: isotherms.

Isolines and isosurfaces

One way to make patterns on a field map easier to see is to draw isolines. **Isolines** are lines that are drawn on a field map to connect all of the points on that map that have the same value. For example, a temperature field map would contain lines connecting points of equal temperature, or isotherms.

Field maps can also be shown in three dimensions using isosurfaces. An isosurface is a three-dimensional surface in which every point on the surface has the same field value. Isosurfaces can help you visualize the field in ways that may not be possible by viewing a two-dimensional map.

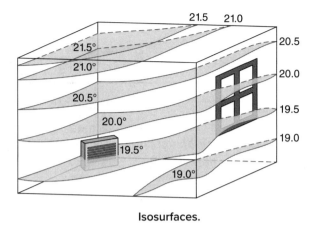

Isosurfaces.

Within a field, the field value changes as you move from place to place. How much the field value changes over a given distance is called its gradient. Gradient can be calculated as follows:

$$\text{Gradient} = \frac{\text{amount of change in the field value}}{\text{distance through which the change occurs}}$$

Fields seldom remain unchanged over time. Field maps show a field at a particular point in time—the time at which its field values were measured. For example, the temperature field over the United States changes quite a bit from July to December. Therefore, field maps need to be updated from time to time in order to show the current field conditions. On weather maps, which show fields that change fast, values are updated as often as once an hour.

PAINLESS TIP

The more closely spaced the isolines on a field map, the steeper the gradient. The steeper the gradient, the faster the field value is changing in that region.

Topographic maps

In Earth science, you will study many shapes and features of Earth's surface, such as mountain ranges, volcanoes, and deep ocean trenches. A map that shows the relative positions and elevations

of the natural and human-made features of a place is called a
topographic map.

PAINLESS TIP

Topographic comes from the Greek words *topos* meaning "place" and
graphien meaning "to write." Topographic literally means "to write
(or draw) a place."

Topographic maps are scale models of a part of Earth's surface that
show its three-dimensional shape on a flat map. Topographic maps
are actually a type of field map in which the field value is the elevation above sea level of points on Earth's surface. The isolines that
connect points of equal elevation are called **contour lines** because
they represent the shape, or contours, of Earth's surface. If you walk
along a contour line, you would not go uphill or downhill. You would
stay at the same elevation above sea level. Each contour line is separated from the next by a specific vertical distance called the **contour
interval**. Note the contour lines on the topographic map on page 19.
Each differs from its neighbor by 100 meters of elevation. Thus,
the contour interval of this map is 100 meters. On actual topographic maps, not every contour line is labeled. This keeps the map
from getting too crowded with numbers. Every fifth contour line is
printed bold and labeled with its elevation.

PAINLESS TIP

To find the contour interval of a topographic map, find two successive
bold, labeled contour lines. Find the difference between the elevations
of the two bold lines and divide by five.

By studying a map's contour lines, you can visualize the shape of the
surface in that place. Note the closely spaced contour lines on the
map. Close spacing means that the elevation changes greatly over
a short distance. In such a place, the ground slopes steeply. If the
contour lines are widely spaced, the ground is gently sloped. The
ground slopes upward on all sides of a hill. On a hill the contour
lines look like concentric rings. The contour lines around a bowl-like

depression would also look like this, so the contour lines around an enclosed depression are printed with hachure marks pointing into the depression.

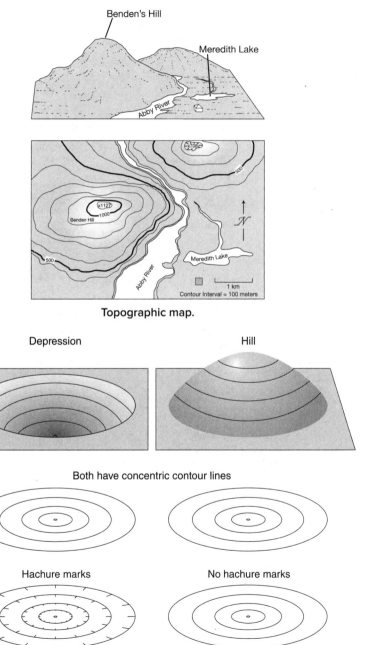

Topographic map.

Depression   Hill

Both have concentric contour lines

Hachure marks   No hachure marks

Depression contours vs. regular contours.

Map symbols

Topographic maps provide a view of the ground as seen from vertically above. Surface features are represented by map symbols. The four most common are blue for bodies of water, black and red for human-made structures, and brown for contour lines and other relief symbols. An extensive list of map symbols can be found below.

Topographic maps.

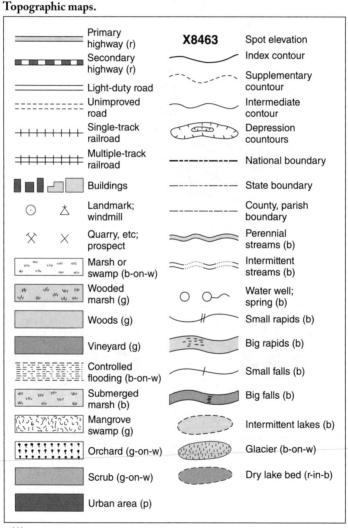

red (r)
blue on white (b-on-w)
green (g)
blue (b)
green on white (g-on-w)
pink (p)
red in blue (r-in-b)

Map scale

A map scale is the ratio between the distance shown on a map and the actual distance on the ground. It is expressed as a ratio such as 1:100,000. Both numbers in the ratio have the same units, but those units could be anything. For example, 1 centimeter on the map equals 100,000 centimeters on the ground. But it also means that 1 inch on the map equals 100,000 inches on the ground. On most topographic maps, map scale is also represented by a graphic scale such as the one shown below. The graphic scale can be used as a ruler to measure distances on the map.

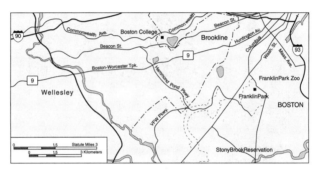

Graphic map scale.

PAINLESS TIP

To measure distances between two places on a map, mark their locations along the straight edge of a piece of scrap paper. Then use the graphic scale as a ruler to measure the distance between the two marks on the scrap paper and determine the actual distance between the two places.

Map profiles

It is often useful to construct a profile from a topographic map. A profile is what a cross-section of the land would look like between two points. The diagram on page 22 shows how a profile can be constructed.

To construct a profile, the two points between which the profile is to be drawn are chosen and a line is drawn connecting them. Then the edge of a piece of paper is placed along the line and the edge of the paper is marked wherever it intersects a contour line.

The paper is then moved to a piece of graph paper on which a vertical scale has been marked to match the contour lines intersected. At each point where a contour line crosses the edge of the paper a line of the appropriate height is drawn on the graph paper. Then the end points of the lines are connected in a smooth line to form the finished profile.

Choosing points (X and y) along which to create a profile

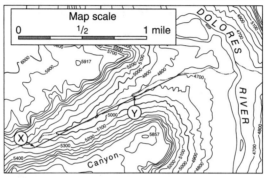

Marking contour line intersections along the line on scrap paper

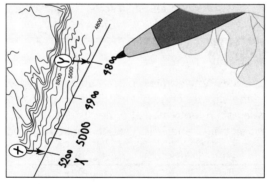

Transferring elevations to the graph paper grid and drawing the profile

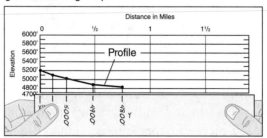

Constructing a map profile.

Tips for reading topographic maps

When topographic maps are drawn or read, the following rules of thumb apply:

- All points on a contour line have the same elevation.
- Every fifth line, called an index line, is generally printed bold and labeled with the line's elevation.
- All contour lines are closed, but they may run off the map.
- Two contour lines of different elevations may not cross each other.
- Contour lines may merge at a cliff or a waterfall.
- The spacing of contour lines indicates the nature of the slope. The closer the contour lines are spaced, the steeper the slope. Even spacing indicates a uniform slope.
- Where contour lines cross a stream, they always form a V, whose apex points up the valley.
- Where contour lines cross a ridge, they often form a V, whose apex points down the valley.

BRAIN TICKLERS   Set # 4

1. A _____ is a region of space in which every point has a measurable property.

    a. map

    b. volume

    c. field

    d. surface

2. Isolines connect points of equal _____ on a field map.

    a. value

    b. location

    c. coordinates

    d. scale

3. The diagram below shows a topographic map of a hill. On which side of the hill does the land have the steepest slope?

a. North

b. South

c. East

d. West

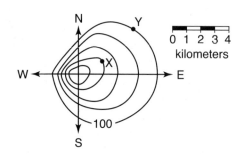

4. The contour lines on the map below represent a hill.

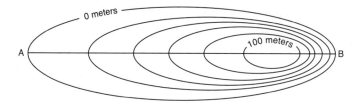

Which hill shape best represents a profile drawn along line AB of the contour map?

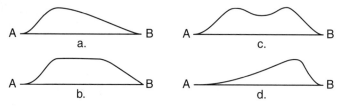

5. Which statement is NOT true about fields?

a. They do not change with time.

b. They are often illustrated by the use of isolines.

c. Any one place in a field has a measurable value at a specific time.

d. Gradients indicate the degree of change from place to place in a field.

6. The field map below shows the temperature (in °C) in a classroom.

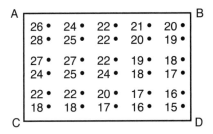

In which diagram do the isolines best represent the temperature field in the classroom?

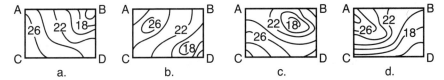

| a. | b. | c. | d. |

7. In the diagram below, the thermometer held 2 meters above the floor shows a temperature of 30°C. The thermometer on the floor shows a temperature of 24°C.

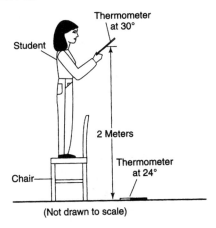

(Not drawn to scale)

What is the temperature gradient between the two thermometers?

a. 2°C/m

b. 3°C/m

c. 4°C/m

d. 6°C/m

8. A topographic map is a two-dimensional model that uses contour lines to represent points of equal

    a. barometric pressure.

    b. elevation above sea level.

    c. temperature gradient.

    d. magnetic force.

(Answers are on page 28.)

Wrapping up

- The rock at Earth's surface forms a nearly continuous shell around Earth called the lithosphere.
- The majority of the lithosphere is covered by a relatively thin layer of water called the hydrosphere.
- Nearly all of the atmosphere is confined to a thin shell surrounding Earth. The atmosphere is a mixture of gases, including nitrogen and oxygen with small amounts of water vapor, carbon dioxide, and other trace gases.
- Earth is a system consisting of parts that interact and are interdependent. Four major interacting components of Earth system are the biosphere, lithosphere, hydrosphere, and atmosphere. Changes occur when different parts of Earth system interact.
- A model is anything that represents the properties of an object or system. Most models are made to scale, meaning that the parts of the model are made in the same proportions as the parts that are in the original.
- Models are used to study processes that cannot be studied directly (e.g., when the real process is too slow, too fast, or too dangerous for direct observation).
- Earth's shape is nearly spherical.
- Positions on Earth's surface are located using the latitude-longitude coordinate system. Latitude lines run east-west, forming circles that are parallel to the equator. Longitude lines run north-south and pass through Earth's poles to form large circles called meridians. Latitude and longitude lines cross

each other at right angles. Latitude is measured in degrees
north or south of the equator. Longitude is measured in degrees east or west of the prime meridian.

- A field is a region of space in which every point has a measurable property. Field maps show how much of something exists
at many locations. Isolines are lines that are drawn on a field
map to connect all of the points on that map that have the
same value. Isolines and isosurfaces make it easier to see patterns in the field properties.

- Topographic maps are field maps of the elevation of points
on Earth's surface. They contain isolines connecting points of
equal elevation above sea level called contour lines.

- Gradient is the amount that the field value changes in a given
distance on a field map.

- A map profile is a picture of what the land would look like
from the side. It is a cross-section of the landscape.

Brain Ticklers—The Answers

Set # 1, page 5

1. L	3. H	5. H	7. H	9. d
2. A	4. B	6. L	8. B	10. b

11. Fish (biosphere) swimming in water (hydrosphere)

    Wind (atmosphere) blows sand (lithosphere) into the ocean (hydrosphere)

    Volcanic eruption (lithosphere) underwater (hydrosphere) kills algae (biosphere) and emits gases that bubble to the surface and enter the air (atmosphere)

Set # 2, page 8

1. Mathematical, visual, physical

2. The real thing is too large, too small, or too dangerous to study directly.

3. a   4. d   5. c

6. You can't see an entire globe at once. A globe would have to be very large to show details like streets.

Set # 3, page 14

1. a

2. Same—both are circles; different—meridians are all the same size, parallels differ in size; different—meridians go through the poles, parallels do not go through poles; meridians run north-south, parallels run east-west.

3. b   4. c   5. b

Set # 4, page 23

1. c	3. d	5. a	7. b
2. a	4. d	6. d	8. b

Chapter 2

The Atmosphere

The atmosphere is a thin layer of gases surrounding the whole planet Earth. The atmosphere protects life on Earth from harmful ultraviolet radiation and from extremes of temperature. Without its atmosphere, Earth's surface would look very different. The Moon has no atmosphere, so rocks from space crash into its surface, forming craters. With no air to support life, the Moon's surface is barren. If Earth had no atmosphere, it would look as cratered and barren as the Moon.

Composition of the Atmosphere

The mixture of many gases and other substances that make up Earth's atmosphere today is known as **air**. Air contains about 78 percent nitrogen, 21 percent oxygen, and 1 percent argon, carbon dioxide, and other gases. Air also contains varying amounts of water vapor, dust, pollen, spores, and chemicals released by industry and microorganisms. The table on page 30 shows the most common gases in dry air. As you can see, nitrogen and oxygen are the most plentiful. Together they make up 99 percent of dry air.

The gases in Earth's atmosphere are important in many ways. Ozone protects living things from the harmful effects of ultraviolet radiation. Green plants need carbon dioxide to make food. Carbon dioxide, ozone, water vapor, methane, and nitrous oxide are greenhouse gases. They absorb heat energy radiated by Earth and in turn radiate it in all directions, including back down toward Earth's surface. Earth's surface is warmer than it would be without an atmosphere because it receives energy from both the Sun and the atmosphere. Without these gases, Earth's surface would be about 33°C (59°F) colder.

Composition of dry air by volume.

Gas	Symbol	Percent by Volume in Dry Air
Nitrogen	N_2	78.08
Oxygen	O_2	20.95
Argon	Ar	0.0934
Carbon dioxide	CO_2	0.0314
Neon	Ne	0.0018
Helium	He	0.0005
Methane	CH_4	0.0002
Krypton	Kr	0.0001
Hydrogen	H_2	0.00005
Xenon	Xe	0.0000087
Ozone	O_3	0.000007
Nitrogen dioxide	NO_2	0.000002

Source: CRC Handbook of Chemistry and Physics.

The amount of water vapor in air varies from place to place and from day to day. In places like a jungle or over a warm ocean, water vapor may make up as much as 4 percent of air. In places like deserts or near the poles, there may be almost no water vapor in the air. Overall, water vapor makes up roughly 0.4 percent of the full atmosphere. But this small percentage of water vapor is the source of every cloud you see in the sky, and every raindrop or snowflake that falls to Earth's surface.

The atmosphere also contains many tiny particles of solids and liquids called **aerosols**. Some of these particles are dust picked up from Earth's surface and carried aloft by winds. Some are shot into the air when volcanoes erupt. Some are crystals of salt left behind when droplets of seawater are blown into the air and evaporate. Others are liquids, such as the tiny droplets of water that make up clouds. Aerosols are important because they can react with gases in the air and scatter or absorb sunlight. They can also affect the health of living things.

CAUTION—Major Mistake Territory!

Water vapor is the gas phase of water. It is *NOT* tiny droplets of liquid water. It is a completely different state of matter than liquid water. Water vapor is a colorless, odorless gas that is less dense than air.

You cannot see water vapor. It is invisible to the human eye. A common mistake is to think that clouds, fog, or even the white mist you see when you breathe out on a cold day are water vapor. These are all made of tiny droplets of *liquid* water, not water vapor. That is why you can see them. The tiny droplets of water look white because they reflect light that strikes them.

Another common mistake is to think that air becomes "heavier" when it contains water vapor. The mistake is in thinking of the water vapor in the air as if it were liquid water. Liquid water is much denser than air. But water vapor is less dense than air. Adding water vapor to air actually makes air less dense because you are displacing heavier molecules of nitrogen and oxygen with lighter molecules of water vapor.

 PAINLESS TIP

When you think of *aerosols*, think of spray cans of deodorant. Some deodorant sprays are liquids; some are powders. The fine mist of liquid droplets or fine particles of solid powder blown out of the can are aerosols.

 BRAIN TICKLERS   Set # 1

1. The mixture of many gases, liquid droplets, and tiny solid particles that make up Earth's present atmosphere is known as _____.

2. Ozone is an important gas in the atmosphere because it protects living things from harmful _____.

3. Without gases such as carbon dioxide, ozone, methane, and nitrogen in the atmosphere, Earth's surface temperatures would be _____.

4. The amount of water vapor in the atmosphere varies from as little as _____ percent to as much as _____ percent.

5. The most accurate description of Earth's atmosphere is that it consists

    a. mostly of oxygen, argon, carbon dioxide, and water vapor.

    b. entirely of ozone, nitrogen, and water vapor.

    c. of gases that cannot be compressed.

    d. of a mixture of gases, liquid droplets, and minute solid particles.

6. Which circle graph best represents the volume of gases in the atmosphere?

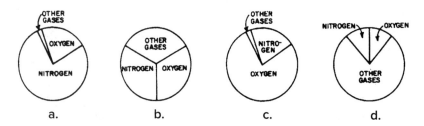

    a.   b.   c.   d.

(Answers are on page 60.)

Structure

The atmosphere extends outward several hundred kilometers from Earth's surface. Most of the air in the atmosphere is held close to Earth's surface by gravity. As altitude increases, the atmosphere becomes thinner and thinner and eventually fades into space. There is no definite boundary between the atmosphere and outer space.

Even though air molecules are not visible, they still have weight and take up space. These tiny molecules whiz to and fro and when they hit a surface, they exert a force. **Air pressure** is the force exerted by the gas molecules that make up air on a given surface area. The molecules in air move in all directions; therefore, air exerts pressure in all directions.

The molecules of gases are widely spaced, so when they are compressed they can fit into a smaller volume. When air is compressed, there are more molecules per unit volume that can exert a force on a surface, so air pressure increases.

At sea level, the weight of the entire overlying atmosphere compresses the air. As you go higher and higher in the atmosphere, there is less and less overlying air. With less and less weight above it, the air becomes less and less compressed. Therefore, air pressure decreases with altitude.

Air pressure is not the only thing that changes with altitude. The composition of the air and the temperature also changes. Earth's atmosphere can be divided into four main layers according to how the temperature changes with altitude. The graphs on page 34 show the physical properties in each layer of the atmosphere.

Troposphere

The **troposphere** is the layer of the atmosphere closest to Earth's surface. The troposphere contains 80 percent of the air in the atmosphere and all of the water vapor. All of Earth's weather, from puffy little clouds to massive hurricanes, occurs in the troposphere.

PAINLESS TIP

The troposphere gets its name from the Greek word *trope*, meaning "to turn" or "overturn." Warm air is less dense than cold air. The warm air in the lower troposphere tends to float upward and the cold air above it tends to sink downward. This layer is named for this overturning of the air.

Much sunlight passes through the atmosphere and warms Earth's surface. The warmed surface heats the air resting on it. Therefore, on average, the lowest part of the troposphere is the warmest and temperatures decrease with altitude. Many mountain tops are still covered by snow in the summer because their altitude is so high that the temperature is below the freezing point of water. The temperature levels off at a frigid –55°C at the top of the troposphere, a boundary called the **tropopause**. At the tropopause, the air stops getting cooler with altitude and is almost completely dry.

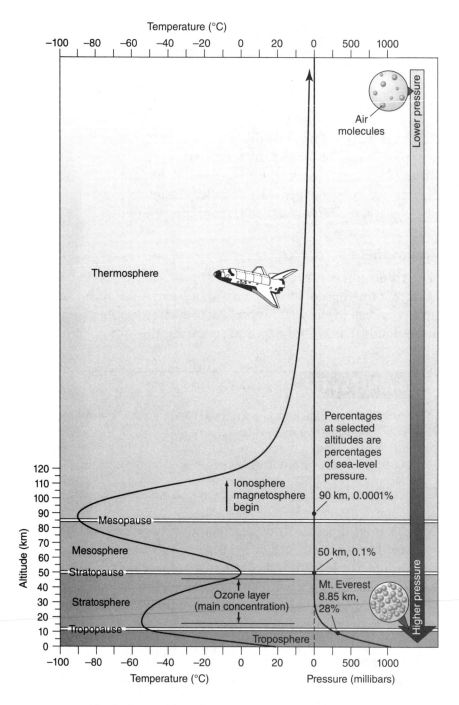

Physical properties of the layers of the atmosphere.

Stratosphere

The layer above the tropopause is the **stratosphere**. In the stratosphere, the air is cold and dry. The temperature remains fairly constant in the lower part of the stratosphere, but then rises because of the **ozone layer**. Ozone gas absorbs ultraviolet (UV) rays from the Sun, which warms the stratosphere. This also prevents many UV rays from reaching Earth's surface and harming living things. Because warmer air is above cooler air in the stratosphere, there is very little mixing or turbulence. At the top of the stratosphere, air temperature reaches a maximum of about 0°C at a boundary called the **stratopause**.

Mesosphere

The **mesosphere** is the layer above the stratopause. Temperature falls steadily in the mesosphere, reaching about −90°C at the **mesopause**. This is the coldest place that is considered part of Earth. The amount of air in the mesosphere is so small that air pressure at the mesopause is only about 1/100,000th of what it is at sea level. At the mesopause, you would have to take 100,000 breaths to take in as much air as one breath at sea level. If you were exposed to air pressure that low, your body would expand and pop like a balloon. However, there is enough air in this layer that it is where most meteors burn up after entering the atmosphere.

Thermosphere

Above the mesopause is the **thermosphere** (or heat sphere). The few molecules that are present in the thermosphere receive huge amounts of energy from the Sun. This causes the layer to warm to well over 100°C—the temperature of boiling water. Air temperature measures how fast the molecules in air are moving, not the total energy they contain. Because air is so thin within the thermosphere, temperatures there cannot be compared with those of the troposphere or stratosphere. Even though the measured temperature is very hot, air in the thermosphere would actually feel cold to us because there are so few molecules that could transfer heat to our skin.

In the lower part of the thermosphere, high-energy radiation from the Sun breaks the few atoms and molecules into charged particles called ions. These ions form the layers of the **ionosphere** that can bounce radio waves from one side of Earth to another. In the thermosphere, the atmosphere gets thinner and thinner until it fades away into outer space.

CAN THE OZONE LAYER BE SAVED?

The ozone layer in the stratosphere protects Earth by absorbing harmful ultraviolet light that can damage the tissues and DNA of living things. If you have ever had a sunburn, your skin tissue has been damaged by ultraviolet light. Damage to the DNA of your skin cells can cause skin cancer.

In the 1970s, scientists noted a steady decrease in the amount of ozone in Earth's atmosphere. Newspapers referred to it as the ozone hole. But the ozone hole is not actually a "hole." It is a region of depleted ozone in the stratosphere over the Polar Regions.

Studies of the stratosphere identified the culprit as chlorofluorocarbons, or CFCs for short. CFCs escape into the atmosphere from spray cans and cooling units like air conditioners and refrigerators, as well as solvents used in dry cleaning and degreasing.

Once in the atmosphere, CFCs make their way to the stratosphere. There, ultraviolet rays break down the CFCs releasing chlorine, which takes part in a number of reactions that destroy ozone. Starting in the late 1980s, many countries agreed to phase out the use of ozone-depleting chemicals. By 2005, humans had cut the amount of CFCs released each year by 90–95%. However, the ozone "hole" still exists because CFCs take a very long time to break down. In 2020, the largest hole ever seen in the Northern Hemisphere appeared over the Arctic and took a long time to close again. However, there is hope. As humans continue to keep CFCs out of the atmosphere, the CFCs in the atmosphere are breaking down. Some models project that by 2040 the ozone "hole" will be mostly recovered.

BRAIN TICKLERS   Set # 2

1. The layers of the atmosphere in the correct order upward from Earth's surface are the _____, _____, _____, and _____.

2. The layer of the atmosphere found directly above the tropopause is the _____.

3. Most of the water vapor in the atmosphere is found in the _____.

4. If you climbed from sea level to the top of a mountain, the weight of the air above you would _____ and the air pressure would _____.

   Base your answers to questions 5 through 7 on the diagram on page 34 and on your knowledge of science.

5. More than 50% of the gas molecules in Earth's atmosphere are found in the troposphere layer. Identify the force responsible for pulling these molecules closer to Earth's surface.

6. The greatest concentration of ozone gas is located at an altitude between 20 and 30 kilometers. Identify the layer of Earth's atmosphere in which the greatest concentration of ozone gas is located.

7. Calculate the temperature difference from the bottom of the mesosphere to the top of the mesosphere as shown on the air temperature line in degrees Celsius.

8. Which graph best represents the relationship between altitude and pressure in the atmosphere?

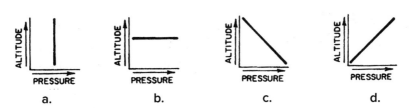

a.   b.   c.   d.

(Answers are on page 60.)

Energy in the Atmosphere

From the frigid howling winds near the poles to gentle tropical breezes near the equator, the atmosphere is constantly moving and changing. The driving force behind all of the changes in the atmosphere is energy from the Sun. Energy from the Sun reaches Earth in the form of electromagnetic waves. But what is an electromagnetic wave?

Electromagnetic waves

All matter is composed of atoms. Every atom contains electrically charged particles such as protons and electrons. These charged particles are surrounded by an electric field. Whenever a charged particle moves, a magnetic force is produced and the particle is then also surrounded by a magnetic field. These two fields—an electric field and a magnetic field—exist at the same time around all electrically charged particles that are moving. Together, they are called an electromagnetic field and extend outward infinitely in all directions around the particle.

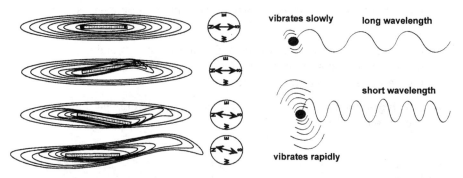

a. Ripples spreading outward in the magnetic field around a magnet.

b. Electromagnetic waves produced by particles vibrating at different rates.

When a particle moves back and forth, its electromagnetic field moves with it. Like a ripple in a pond, this movement spreads out through the field in the form of a wave. Disturbances in electromagnetic fields around particles can travel through space as they move outward, or radiate, through the field. In this way, energy is transferred from the Sun to Earth through the emptiness of space in the form of a wave. Electromagnetic waves move in straight lines and travel at the speed of light.

The electromagnetic spectrum

Elecromagnetic waves are **transverse waves**. In a transverse wave, the wave vibrates at right angles to the direction in which it is moving. The distance from one point on a wave to the corresponding point on the next wave is called the **wavelength**. The height of the wave is called its **amplitude**. The number of waves that occur in a given time is called **frequency**.

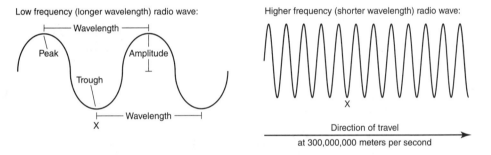

Transverse wave showing wavelength, amplitude, and frequency.

The way in which the source particle moves affects the way in which the wave it forms moves through the electromagnetic field. The faster the source particle moves back and forth, the shorter the wavelength, or distance between "ripples" in the electromagnetic field. The shorter the wavelength, the more waves occur in a given time and the higher the frequency. Every speed at which a source particle moves back and forth will produce a wave of a different wavelength and frequency.

Electromagnetic waves are classified by their wavelength. The diagram on page 40 shows the **electromagnetic spectrum**, a chart in which the different types of electromagnetic waves are arranged in order from longest to shortest wavelength. If a wave has a wavelength, or distance between "ripples," of 1 centimeter, it is a radio wave. If it has a wavelength of 650 nanometers, then it is red visible light. Visible light is only a narrow band in the middle of the spectrum.

All of the other types of electromagnetic waves are not visible to the human eye. The Sun emits many electromagnetic waves with many different wavelengths because it contains particles moving at many different speeds. However, most of the waves coming from the Sun are in the visible light and infrared range.

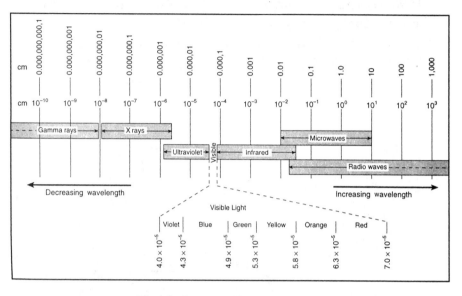

The electromagnetic spectrum.

To understand how electromagnetic waves from the Sun interact with Earth, you need to know that all matter is made of tiny particles called molecules. These tiny molecules are always moving. The energy of movement of the molecules in a substance is called **thermal energy**. The more energy the molecules have, the faster they move. **Temperature** is a measure of the average speed of the molecules in a substance. The higher the temperature, the faster the molecules move; thus, temperature can be used as a measure of thermal energy. We sense thermal energy as **heat** when it is transferred from one substance to another.

If you have ever been struck by a wave at the beach, you have felt the energy in a water wave. An electromagnetic wave also contains energy. It can exert forces on matter with which it interacts. When an electromagnetic wave from the Sun reaches Earth, the moving field interacts with the electromagnetic fields surrounding molecules of matter. When electromagnetic waves interact with matter they can be:

- **transmitted**, which means they pass through the matter;
- **reflected**, which means they are thrown back or returned from the surface of the matter;
- **refracted**, which means their path is bent as it passes through the matter;
- **scattered**, which means reflected or refracted so that they are dispersed or go in different directions; or
- **absorbed**, meaning the wave energy is taken into the substance.

When electromagnetic waves radiating from the Sun reach Earth's atmosphere, about one-third are reflected back into space, mostly by clouds. As the remainder pass through the atmosphere, some are absorbed by gas molecules. Some are refracted as they cross boundaries between layers of different densities. Some are scattered by dust or aerosols in the air. The solar radiation that is transmitted and reaches Earth's surface is called **insolation**, short for *in*coming *sola*r radi*ation*. Most of the electromagnetic waves that make up insolation are visible light waves. Insolation is either reflected or absorbed by Earth's surface. These interactions are summarized in the chart on page 42.

Factors affecting insolation

Earth's surface is made up of many different kinds of substances, including many kinds of rocks and soils, water, ice, and vegetation. All interact differently with the insolation that strikes them.

Color and texture

A ball is more likely to bounce off a concrete wall than a pillow. Whether a substance is more likely to absorb or reflect insolation depends on its molecular structure. The color of a substance is a fairly good indicator of the amount of insolation it absorbs or reflects. The more light a substance reflects, the lighter its color. The more light a substance absorbs, the darker its color. In general, dark-colored substances like black soil absorb more insolation than light-colored substances like sand. The more insolation a substance absorbs, the more it is heated.

	Incoming Solar Radiation	
	Average Gained by Earth (%)	Average Lost Back to Space (%)
Reflection from clouds to space		21
Diffuse reflection by dust and aerosols (scattering) to space		5
Direct reflection by Earth's surface		6
Absorbed by clouds	3	
Absorbed by dust, water vapor, CO_2, and other gas molecules	15	
Absorbed by Earth's surface	50	
Total	68	32

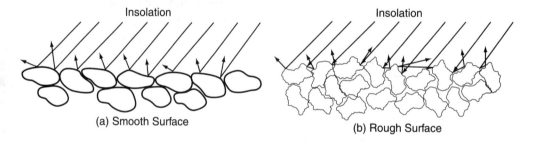

Insolation striking surfaces with rough and smooth textures.

The bumps and pits on a surface with a rough texture can cause insolation to reflect at angles that can cause it to hit the surface several times before leaving. Each time insolation hits the surface, a little more of its energy is absorbed. Therefore, rough surfaces tend to absorb more insolation than smooth surfaces.

Specific heat

When different substances absorb the same amount of insolation, they do not all change temperature by the same number of degrees. **Specific heat** is a measure of how much energy is needed to cause a temperature change of 1°C in 1 gram of a substance.

The table below shows the specific heats of different substances and the temperature change that occurs in each with the addition of the same amount of heat. Notice that the temperature of water changes less than that of rocks like granite or basalt. Therefore, water doesn't get as hot as nearby land when exposed to the same insolation.

Specific heat of common substances.

Substance		Specific Heat (joules/gram · °C)	Temperature Increase if 1 Joule of Heat is Added to 1 Gram of the Substance (°C)
Water	Solid	2.11	0.47
	Liquid	4.18	0.24
	Gas	2.0	0.5
Dry air		1.01	1.0
Basalt		0.84	1.2
Granite		0.79	1.3
Marble		0.88	1.1
Iron		0.45	2.2
Copper		0.38	2.6
Lead		0.13	7.7

Angle of insolation

Since Earth is a sphere, sunlight does not strike all places on its surface at the same angle. Near the equator, the Sun's rays strike the surface most directly (nearest 90°), but near the poles they strike at more of a slant. As you can see below, the direct rays are concentrated in a smaller area, whereas the slanted rays are spread out over a wider area. This means the energy of the direct rays is concentrated in a smaller area and heats the surface more than rays that are spread out. That is why it is hotter near the equator than near the poles.

The angle of the Sun's rays also changes during the course of each day. The Sun's rays strike most directly around noon and are most slanted at dawn and sunset. That is why the warmest part of each day usually occurs in the middle of the day—the more direct noontime rays heat Earth's surface more than the slanting rays of early morning or evening.

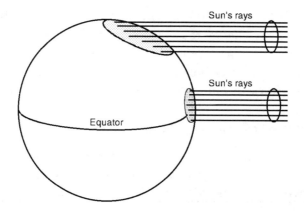

Sunlight striking Earth's surface near the equator and near the poles.

Duration of insolation

The **duration of insolation** is the length of time that insolation is received each day. Earth's surface only receives insolation during the daylight hours. Therefore, the greater the number of daylight hours, the more insolation Earth's surface receives.

BRAIN TICKLERS   Set # 3

1. Earth's surface temperatures are due mainly to energy received from the _____.

2. Electromagnetic energy from the Sun that reaches Earth's surface is called _____.

3. Most of the electromagnetic waves that pass through the atmosphere and reach Earth's surface are in the form of
_____.

4. A surface will most effectively reflect insolation if it has a
_____ texture and a _____ color.

5. Bodies of water warm up slower than land areas because water has a
_____ specific heat than land.

6. The diagram below shows light rays striking three different surfaces, A, B, and C.

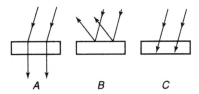

Which statement best describes what happened to the light rays after striking each surface?

a. The light was reflected in A, absorbed in B, and transmitted in C.

b. The light was transmitted in A, reflected in B, and absorbed in C.

c. The light was reflected in A, transmitted in B, and absorbed in C.

d. The light was absorbed in A, reflected in B, and absorbed in C.

7. Each of the sunbeams in the diagrams below contains the same amount of electromagnetic energy, and each sunbeam is striking the same type of surface. Which surface is receiving the greatest amount of energy per unit area where the sunbeam strikes the surface?

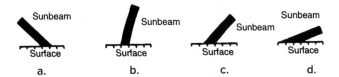

Base your answers to questions 8 and 9 on the diagram below that shows part of the electromagnetic spectrum.

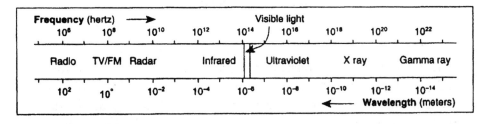

8. List the following three types of electromagnetic energy in order from shortest wavelength to longest wavelength: infrared, X-rays, red visible light.

9. Which type of electromagnetic energy reflected by the Moon is most easily seen by a person on Earth?

10. List *two* forms of electromagnetic energy that have shorter wavelengths than visible light.

(Answers are on page 60.)

Energy Transfer in the Atmosphere

As you can see, the atmosphere does not absorb much solar energy directly. How, then, is the atmosphere heated? Most insolation passes right through the atmosphere to Earth's surface where it is absorbed and changed into thermal energy. The atmosphere is said to be transparent to this radiation. Let us now consider how thermal energy moves from Earth's surface into the atmosphere and, once there, how it moves within the atmosphere.

Conduction

Solar radiation absorbed by Earth's surface causes the molecules at the surface to move more rapidly. Earth's surface is in continuous contact with the base of the atmosphere. Therefore, faster-moving surface molecules collide with slower-moving molecules in the air. During these collisions, thermal energy is transferred from Earth's surface molecules to the gas molecules of the air. This causes the gas molecules to move faster, increasing the temperature of the air. This kind of energy transfer, in which thermal energy moves from molecule to molecule by collisions, is known as **conduction**. Conduction is important because it is the main way in which thermal energy moves from Earth's surface to the atmosphere.

Convection

The gas molecules in air are more widely spaced than those of the solids and liquids that make up Earth's surface. As a result, collisions are less likely and thermal energy spreads very slowly by conduction in the atmosphere. As thermal energy builds up in the air touching the ground faster than it can be conducted away, the air becomes warmer and warmer.

As the air warms, its molecules move faster and spread farther apart. This causes the air to expand and become less dense than the surrounding air. The warm air floats upward, or rises, and cooler air moves in to take its place, a type of movement known as **convection**.

During convection, the rising warm air carries the thermal energy it contains upward along with it. In this way, convection transfers thermal energy from place to place within the atmosphere. Convection is important because it is the way that energy that was conducted into the atmosphere from Earth's surface is then carried throughout the atmosphere. It is also important because it is what causes air in the atmosphere to move from place to place.

Radiation

Energy from the Sun reaches Earth by radiation, the release and transfer of energy in the form of electromagnetic waves. As described earlier, an electromagnetic wave does not need a substance through which to move. The energy in an electromagnetic wave can radiate from its source across empty space. This is the method by which the Sun's energy gets to Earth (and other solar system objects) through space. It is also the method by which it passes through the atmosphere to reach Earth's surface.

The greenhouse effect

You already know that all moving atoms and molecules emit electromagnetic radiation, and the wavelength emitted depends on how fast they are moving. However, since Earth and its atmosphere are both much cooler than the Sun, the energy they radiate has much longer wavelengths. Most of the energy radiated by Earth's surface is infrared radiation. This is important, because the atmosphere is transparent to most of the Sun's short-wavelength radiation. But it is not as transparent to infrared radiation. Ozone, carbon dioxide, and water vapor in the atmosphere absorb some of Earth's infrared radiation and cause the atmosphere to become warmer. The atmosphere then radiates more energy back toward Earth. Therefore, Earth receives energy from both the Sun and from the atmosphere, a phenomenon known as the **greenhouse effect**.

It is called the greenhouse effect because just as glass surrounds a greenhouse, the atmosphere surrounds Earth. The greenhouse effect keeps Earth warmer than it would be otherwise because Earth receives energy from two sources: the Sun and the atmosphere.

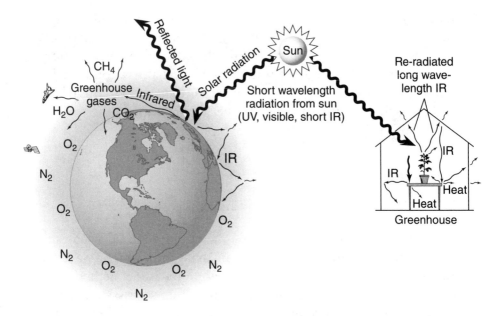

The greenhouse effect showing short-wave solar radiation passing through atmosphere and long-wave terrestrial radiation being absorbed.

PAINLESS TIP

When you think about the greenhouse effect, think about what happens to a car parked outside on a sunny day. Solar radiation comes through the transparent windows and is absorbed by the surfaces inside. Heat radiation given off by these surfaces is blocked by the glass and trapped in the car.

BRAIN TICKLERS   Set # 4

1. The Sun's energy travels through space by the process of
   _____.

2. The transfer of energy from Earth's surface to the atmosphere by molecular collisions is called _____.

3. Air flows from one location in the atmosphere to another by the process of _____.

Base your answers to questions 4–6 on the diagram below. The water was heated for several minutes.

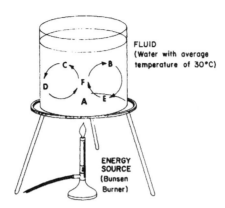

4. The type of energy transfer indicated by the arrows in the diagram is called _____.

5. The density of the water in the beaker is least at point

    a. A.

    b. B.

    c. C.

    d. D.

6. As water moves from point A to point F to point B, the water's temperature will generally _____.

Base your answers to questions 7–9 on the diagram and data below. The diagram represents a closed glass greenhouse. The data table shows the air temperatures inside and outside the greenhouse from 6 A.M. to 6 P.M. on a sunny day.

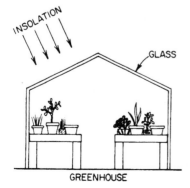

AIR TEMPERATURE

Time	Average Outside Temperature	Average Inside Temperature
6 A.M.	10°C	13°C
8 A.M.	11°C	14°C
10 A.M.	12°C	16°C
12 noon	15°C	20°C
2 P.M.	19°C	25°C
4 P.M.	17°C	24°C
6 P.M.	15°C	23°C

7. The highest temperature was recorded at

    a. 12 noon outside the greenhouse.

    b. 2 P.M. outside the greenhouse.

    c. 12 noon inside the greenhouse.

    d. 2 P.M. inside the greenhouse.

8. By which process does air circulate inside the greenhouse due to differences in air temperature and air density?

    a. Absorption

    b. Radiation

    c. Convection

    d. Conduction

9. Which model best represents why the air in the greenhouse becomes warmer than the outside air as a result of insolation from the Sun?

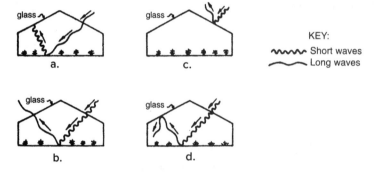

KEY:
〰️ Short waves
〜 Long waves

(Answers are on page 60.)

Circulation in the Atmosphere

Unequal heating of Earth's surface results in unequal heating of the atmosphere. As you know, when air is heated, its molecules move faster and spread farther apart. This causes the warm air to become less dense than the surrounding air and the warm air rises. In cooler air, the molecules move slower and are closer together. Therefore, the cooler air is denser and sinks. As air rises and sinks, surrounding air moves in from the sides to replace it, setting up a circular pattern of convection known as a **convection cell**. In a convection cell, air circulates in the atmosphere due to density differences.

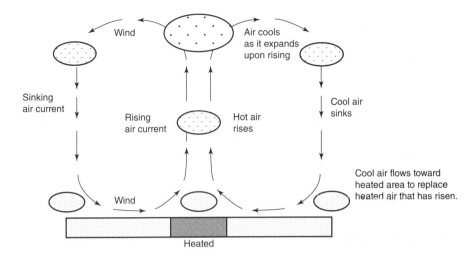

Heated air rises and expands, then the air cools and sinks, forming a circular convection current, or convection cell.

Winds and air currents

The circular movement of air in a convection cell involves both vertical and horizontal movements. Vertical movements of rising and sinking air are called **air currents**. When sinking air reaches Earth's surface, it spreads out horizontally. When rising air expands it also spreads out horizontally. Horizontal movements of air parallel to Earth's surface are called **winds**.

Winds are caused by differences in air pressure. Rising warm air is less dense and exerts less pressure than sinking cool air. It is easy to understand how differences in air pressure cause winds if you think of two people pushing against each other. The person pushing harder will advance against the person pushing with less force. In the same way, air exerting high pressure will advance against air exerting lower pressure. The movement of the advancing air creates wind. Winds blow from regions of high pressure toward regions of low pressure. The greater the difference in pressure, the faster the high-pressure air advances and the greater the wind speed.

PAINLESS TIP

Winds blow from high to low.

Winds are described by their speed and direction. Wind direction is the direction from which a wind blows. Just as a person who comes from the south is called a southerner, a wind that blows from the south is called a south wind. Wind speed is simply the speed at which the air is moving, usually expressed in kilometers per hour.

Local convection and winds

Earth's surface is made up of many different substances. These substances do not all heat up at the same rate when they absorb insolation. Unequal heating on a small scale produces small convection cells that give rise to local winds. Land and sea breezes are good examples of this.

During the day, adjacent land and water receive the same amount of insolation. But the land increases in temperature more than water because water has a higher specific heat than soil. As a result, air over land will have a higher daytime temperature than air over nearby water. Since warm air exerts less pressure than cool air, the pressure over the land will be lower and the pressure over the water will be higher. This results in the movement of air from the water toward the land. Since a wind is named for where it is blowing from, this type of wind is called a **sea breeze**.

At night though, land cools off faster than the water. The air over the land becomes cooler than the air over the water. The air pressure is then greater over the land than over the water and air moves from the land toward the water. This type of wind is called a **land breeze** because it comes from the land.

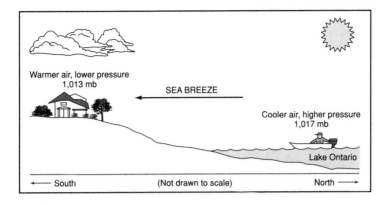

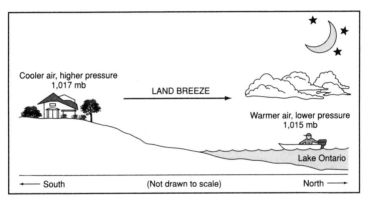

Global convection cells

Unequal heating on a global scale produces global convection cells. As you know, insolation is more direct near the equator and less direct near the poles. Therefore, Earth's surface is warmer near the equator than it is near the poles. This temperature difference causes three sets of global-size convection cells to form. See figure on the following page.

Regions near the equator are not only warm; they are also mostly covered with water so the air also becomes very moist. The warm, moist air at the equator is less dense and exerts less pressure than the cooler, drier surrounding air. The result is a zone of low-pressure air over the equator. Because this zone circles the globe like a belt, it is also called a low-pressure belt.

The higher-pressure surrounding air pushes inward toward the equator, displacing the warm, moist air at the equator upward. As the air is forced upward it expands outward and cools, causing water vapor to condense into liquid water and fall back to Earth. The end result is cooler, drier air that is now denser than the air beneath it. This denser air sinks back toward the surface creating high-pressure belts at about 30°N and 30°S latitude. Together, the rising warm, moist air and sinking cool, dry air form a convection cell on either side of the equator.

Similar convection cells form over the poles. But they begin with cold air that sinks over the poles creating a high-pressure belt. When the sinking cold air reaches Earth's surface, it spreads outward in all directions. As the air moves away from the poles, it warms and rises upward creating a low-pressure belt at about 60°N and 60°S latitude.

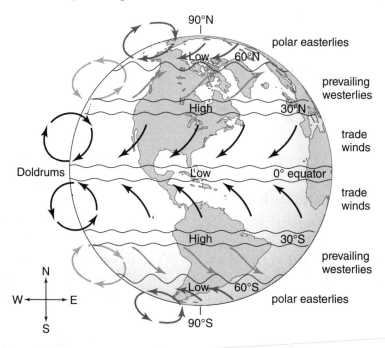

The three-cell theory of atmospheric convection—global, wind, and pressure belts.

Between the polar and equatorial convection cells is a third cell that is set in motion by the winds moving from the high-pressure belts at 30° latitude toward the low-pressure belts at 60° latitude. This three-cell theory of global atmospheric convection explains how the atmosphere distributes solar energy over the whole Earth.

The Coriolis effect

Earlier, you learned that winds blow from regions of high pressure toward regions of low pressure. Notice that global winds do not blow in a straight path between the global high- and low-pressure belts. Instead, the winds curve to the right in the Northern Hemisphere and to the left in the Southern Hemisphere. This does not only happen to winds. It also happens to ocean currents, airplanes, rockets, or any other matter moving over Earth's surface. Moving objects appear to curve as they move over Earth's surface because Earth is rotating while the matter is moving over it. The apparent change in direction of a moving object due to Earth's rotation is called the **Coriolis effect**.

To better understand the Coriolis effect, look at the figure below. The turntable in figure a is not moving. A ball covered with ink is rolled across the turntable, striking the target. When you look at the ink trail, you see that the ball's path is a straight line. In figure b, the turntable is rotating. Again the ball is rolled across the turntable. It rolls away from you in a straight line. But this time, when you look at the ink trail, you see that the ball's path is curved. The ball rolls off the turntable to the right of the target. The Coriolis effect due to the turntable's rotation causes the ball to make a curved path instead of a straight one.

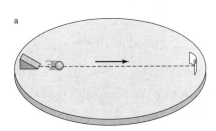

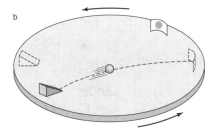

(a) Ink-covered ball shoots across a turntable that is not rotating. The ball leaves a straight path and hits the target.

(b) Ink-covered ball shoots across a rotating turntable. The ball leaves a curved path and misses the target.

> ### 👆 PAINLESS TIP
>
> Remember that there is an *R Right* in the middle of *noRth*, and that *South*paws are *left*-handed.
>
> The Coriolis effect causes moving objects to appear to curve to the *right* in the *North*ern Hemisphere and to the *left* in the *South*ern Hemisphere.

Global wind belts

Look at the diagram on page 54. Air moving from the high-pressure belts toward the low-pressure belts creates a series of global wind belts. Because of the Coriolis effect, these global wind belts curve to the right in the Northern Hemisphere and to the left in the Southern Hemisphere. The wind belts blowing away from the high pressure at the poles are called the **polar easterlies**. They are called easterlies because they blow from the east. The wind belts blowing toward the equator are called the **trade winds**. They are called the trade winds because they were used by captains of sailing ships to cross the world's oceans and establish trade routes. Now look at the air sinking to create the high-pressure belts at 30°N and 30°S latitude. When the sinking air reaches the ground, it spreads outward. Part flows toward the equator becoming the trade winds. But part also flows toward the low-pressure belts at 60°N and 60°S. The wind belts in the middle latitudes are called the **prevailing westerlies**. They are called westerlies because they blow from the west. The prevailing westerlies steer weather systems eastward as they move across the United States.

If you were standing at 30°N latitude looking north, your right hand would point east. Therefore, the prevailing westerlies curve to the right due to the Coriolis effect and travel from west to east. If you were standing at 30°S looking south, your left hand would point east. The prevailing westerlies would therefore curve to the left due to the Coriolis effect, but still end up traveling from west to east.

BRAIN TICKLERS   Set # 5

1. The primary cause of winds is the _____ heating of Earth's atmosphere.

2. Wind moves from regions of _____ pressure toward regions of _____ pressure.

3. Compared to polar areas, why are equatorial areas of equal size heated much more intensely by the Sun?

   a. The Sun's rays are more nearly perpendicular at the equator than at the poles.

   b. Areas near the equator contain more water than polar areas.

   c. More hours of daylight occur at the equator than at the poles.

   d. The equatorial areas are nearer to the Sun than the polar areas are.

4. The diagram right shows a city on a sunny afternoon.

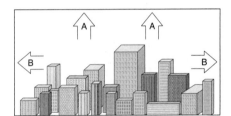

The air in the city is being

   a. heated and will move in the direction shown by the arrows labeled A.

   b. heated and will move in the directions shown by the arrows labeled B.

   c. cooled and will move in the direction shown by the arrows labeled A.

   d. cooled and will move in the directions shown by the arrows labeled B.

5. Adjacent water and land masses are of equal temperature at sunrise. They are heated by the morning sun on a clear, calm day. After a few hours, a surface wind develops. Which diagram best represents this wind's direction?

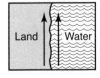

a.   b.   c.   d.

Base your answers to questions 6–8 on the diagram below.

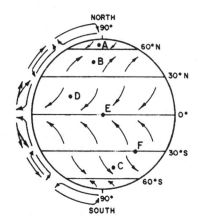

6. The curving paths of the surface winds shown in the diagram are caused by Earth's

   a. gravitational field.   **c.** rotation.

   b. magnetic field.   **d.** revolution.

7. Which location is experiencing a southwest planetary wind?

   a. A

   b. B

   c. C

   d. D

8. The arrows in the diagram represent energy transfer by which process?

   a. Conduction   **c.** Convection

   b. Radiation   **d.** Absorption

(Answers are on page 60.)

Wrapping up

- The atmosphere is made up of a mixture of gases and other substances known as air.

- Air pressure is the weight of the air on a given area. Air pressure decreases with altitude.

- Earth's atmosphere can be divided into four main layers according to how the temperature changes with altitude: troposphere, stratosphere, mesosphere, and thermosphere.

- Energy from the Sun reaches Earth in the form of electromagnetic waves. Electromagnetic waves are classified by their wavelength in the electromagnetic spectrum.

- When electromagnetic waves interact with matter they can be transmitted, reflected, refracted, scattered, or absorbed.

- Insolation is the solar radiation that reaches Earth's surface after passing through the atmosphere.

- The heating of Earth's surface by insolation depends on the angle and duration of insolation, and the color, texture, and composition of the surface.

- Thermal energy can be transferred by conduction, convection, or radiation.

- Winds blow from regions of high pressure toward regions of low pressure.

- The Coriolis effect is the apparent change in direction of a moving object due to Earth's rotation.

Brain Ticklers—The Answers

Set # 1, page 31
1. air
2. ultraviolet radiation
3. colder
4. 0; 4
5. d
6. a

Set # 2, page 37
1. troposphere; stratosphere; mesosphere; thermosphere
2. stratosphere
3. troposphere
4. decrease; decrease
5. gravity *or* gravitational attraction
6. stratosphere
7. 90°C *or* –90°C
8. c

Set # 3, page 44
1. Sun
2. insolation
3. visible light
4. smooth; light
5. higher
6. b
7. b
8. X-rays, red visible light, infrared
9. visible light
10. Any *two* of the following: ultraviolet light (UV) or X-rays or gamma rays.

Set # 4, page 48
1. radiation
2. conduction
3. convection
4. convection
5. a
6. decrease
7. d
8. c
9. d

Set # 5, page 57
1. unequal
2. high; low
3. a
4. a
5. a
6. c
7. b
8. c

Chapter 3

Weather and Climate

Weather describes the conditions in the atmosphere at a given place and time. The major cause of weather is circulation of the air in the atmosphere. This circulation causes the air at any one place to constantly be moved away and replaced by air from a different place—causing the weather to change.

Weather Variable

The constantly changing weather is described in terms of the characteristics of the atmosphere that change, or **weather variables**. The weather variables used to describe the atmosphere include air temperature, air pressure, humidity, wind speed and direction, clouds, and precipitation. Weather instruments are used to measure weather variables. See table on page 63.

Air temperature

Air temperature is a measure of the average speed of the molecules in a substance. It is related to the amount of thermal energy present in the atmosphere at a given place. This energy comes from solar radiation that is absorbed by Earth's surface. This energy is then transferred into the air as it passes over Earth's surface.

Air temperature is measured with different types of thermometers. A **thermometer** is an instrument that measures temperature. A bulb thermometer has a glass tube with a bulb at the end filled with a liquid, usually mercury or colored alcohol. It measures temperature by the expansion and contraction of the liquid. The more the liquid heats up, the more it will expand. By measuring the height of the liquid in the tube, you indirectly measure its temperature.

There are different scales used to measure temperature. When a thermometer is placed in water that is changing phase, the liquid inside it stops expanding for a time because water does not change temperature while changing phase. Thus, the freezing point and boiling point of water make good reference points for measuring the expansion of the liquid in the thermometer. The thermometer is placed in freezing water and the height of the liquid is marked. Then it is placed in boiling water and the height of the liquid is marked. The space between the freezing and boiling point marks is then divided into equal parts, called degrees.

In the **Celsius scale**, the space is divided into 100 equal parts and a value of 0°C is assigned to the freezing point and 100°C to the boiling point. In the **Fahrenheit scale,** the space is divided into 180 equal parts and a value of 32°F is assigned to the freezing point and 212°F to the boiling point of water. The Fahrenheit scale is commonly used to measure air temperature in the United States. Most other countries (and scientists) use the Celsius scale.

Air pressure

Air pressure is the force exerted by air molecules when they collide with a surface. Air pressure is due to the total weight of the air molecules above a given unit of area. The greater the weight of the overlying air, the closer together the molecules are squeezed. This means that more molecules will collide with a surface and the air pressure will be higher.

A number of different units can be used to describe air pressure. The **atmosphere** (atm) is the average pressure exerted by the weight of Earth's atmosphere at sea level. A pressure of 1 atmosphere is about the same as having the weight of a 1 kilogram mass pressing against every square centimeter of a surface (or about 14.7 pounds per square inch). In meteorology, air pressure is measured in units called **millibars** (mb). At sea level, a normal air pressure of 1 atmosphere is defined as 1013.2 millibars.

Weather variables, weather instruments, and how the instruments work.

Weather Factor	Instrument		How It Works
Air temperature	Thermometer		Liquid expands when heated and rises in tube.
Air pressure	Barometer		Air exerts pressure on pool of mercury, forcing it up tube.
Humidity	Sling psychrometer		Evaporation from wet-bulb thermometer causes cooling. The drier the air, the more evaporation, the more cooling, and the greater the difference between wet-bulb and dry-bulb temperatures.
Wind speed	Anemometer		Wind exerts force on cups, causing anemometer to spin. The higher the wind speed, the faster it spins.
Wind direction	Wind vane		Wind exerts more force on tail than on tip of arrow, causing vane to swing and point in direction from which wind is blowing.
Precipitation	Rain gauge		Measures volume of liquid precipitation collected.

Measuring air pressure

Air pressure is measured with different types of **barometers**. In a **mercury barometer**, the air pressure outside forces liquid mercury up inside of a tube with no air in it. Pressure is measured by how far up the tube the mercury is pushed. This is why pressure is often described as "inches or millimeters of mercury."

Aneroid means no air. An **aneroid barometer** is a can with no air inside of it. However, it does contain a spring scale. Air pressure on the sides of the can compresses the spring. As air pressure increases or decreases, the spring compresses and expands and the scale records the force exerted on the can.

> ## PAINLESS TIP
>
> Barometer is from the Greek "baro" meaning "weight" and "meter" or "to measure."

Factors affecting air pressure

Air pressure changes with air temperature. When air heats up, its molecules move faster. You might think that this would increase air pressure because fast-moving molecules hit a surface harder than slow-moving molecules. But that is not the case. When air heats up, its molecules also spread farther apart. This means that fewer molecules will hit the surface. The number of molecules hitting the surface has more of an effect than their speed, so *increasing temperature decreases air pressure.*

Air pressure also changes as water vapor enters and leaves the air. When water vapor enters the air, it displaces molecules of other gases, such as nitrogen and oxygen. Water vapor molecules have less mass than molecules of nitrogen or oxygen. The less a molecule's mass, the less force it exerts when it hits a surface. Therefore, *increasing the amount of water vapor in air decreases air pressure.* Displacing nitrogen and oxygen molecules with water vapor molecules that have less mass also decreases the mass of a given volume of air; that is, it decreases its density. Thus, moist air is also less dense than dry air.

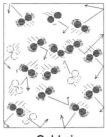 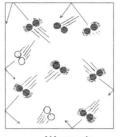

Cold air   Warm air

Increasing air temperature causes air to expand,
decreasing air pressure.

 **PAINLESS TIP**

Think about adding water vapor to air as displacing golf balls (nitrogen or oxygen) with ping pong balls (water). The ping pong balls are lighter and would also hit a surface with less force than a golf ball moving at the same speed.

Humidity

Humidity refers to the amount of water vapor in the air. Humidity is not droplets of liquid water suspended in the air; it is water vapor—a colorless, odorless gas. Most of the water vapor comes from the hydrosphere, but some is given off by plants and animals. Where there are large areas of water, large amounts of water vapor enter the air as water evaporates.

Wherever air comes in contact with liquid water, there is a constant two-way flow of water molecules. Water molecules that gain enough energy from sunlight or other sources *leave* the liquid and enter the air as a gas, or **evaporate**. At the same time, water vapor molecules in the air that arrive at the surface of the liquid may be trapped there and *return* to the liquid, or **condense**. Water vapor may also condense when molecules that have lost energy come in contact with the surfaces of dust or water droplets in the air, stick to their surfaces, and are trapped. When the number of molecules leaving and returning is in balance, we say the water vapor in the air is in **equilibrium**.

Temperature affects this balance. At high temperatures, more molecules evaporate than condense. Thus, more water vapor enters the atmosphere before balance is reached. As the temperature decreases, molecules have less energy, so more return than leave. Thus, the air will contain less water vapor when balance is reached. Therefore, *for any given temperature, there is a specific amount of water vapor the air will contain when it is in equilibrium.*

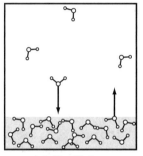 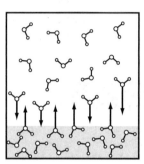

(a) Equilibrium
at low temperature

(b) Equilibrium
at high temperature

Evaporation and condensation at equilibrium.

PAINLESS TIP

Think of equilibrium at different temperatures like juggling balls in the air. The faster the juggler throws balls into the air (higher temperature), the more balls will be in the air at any given time. At any given rate of throwing balls in the air, the number of balls going up will be the same as the number of balls coming down; that is, they will be in equilibrium.

Relative humidity

Relative humidity is the current amount of water vapor in the air compared to the amount of water vapor in the air at equilibrium. For example, a relative humidity of 75 percent means the air contains 75 percent of the water vapor that it would contain if it were in equilibrium at the current temperature. The higher the relative humidity, the more uncomfortable people feel because less moisture evaporates from their skin before equilibrium is reached. Since less water evaporates from their skin, they feel wet and "sticky."

Relative humidity is also a good indicator of how likely it is that the weather will change. High relative humidity means the water vapor

in the air is close to its equilibrium amount. A decrease in temperature could drop the equilibrium amount *below* the amount in the air.
The excess water vapor would then condense out of the air, forming
clouds, fog, or precipitation.

Dew point

The temperature at which the water vapor in a given part of the air
is in equilibrium is called the **dew point**. When the air temperature
reaches the dew point, the water vapor in the air begins to condense
into droplets of liquid water. The condensed water is called dew.

Measuring humidity

Humidity is measured with a hygrometer or a psychrometer. A
hygrometer measures humidity by its effect on human hair; it has
strands of human hair attached to a pointer. Human hair lengthens
slightly as humidity increases, changing the position of the pointer.

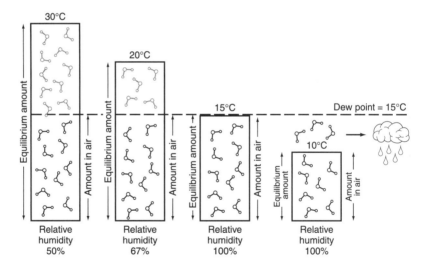

Condensation at the dew point temperature.

A **psychrometer** measures humidity by the cooling effect of evaporation. Evaporation causes cooling because the fastest moving molecules leave the liquid. The slower molecules left behind then have
a lower average speed, or lower temperature. A psychrometer has
two thermometers, one whose bulb is kept dry and another whose
bulb is kept wet with a cloth wick soaked in water. Evaporation from
the wet-bulb thermometer lowers its temperature. The dry-bulb

thermometer remains unchanged at the temperature of the air. The lower the relative humidity, the more water evaporates from the wet bulb and the lower its temperature drops. Therefore, the difference between the temperature of the wet- and dry-bulb thermometers is directly related to the relative humidity of the air.

PAINLESS TIP

When comparing the temperatures of the wet-bulb thermometer and the dry-bulb thermometer in a psychrometer, the bigger the **difference**, the **drier** the air.

Wind

As you learned in Chapter 2, air in the atmosphere circulates due to density differences caused by unequal heating. The vertical movements of rising and sinking air are called air currents. When sinking air reaches Earth's surface, it spreads out horizontally. When rising air expands it also spreads out horizontally. Horizontal movements of air parallel to Earth's surface are called winds. Wind is described by both its speed and its direction.

Measuring wind speed and direction

Wind speed is measured with an **anemometer**. An anemometer has several large cups mounted on a shaft. The cups catch the wind and make the shaft spin. The speed at which the axis spins varies with wind speed. Wind speed is measured in knots, a nautical measure of speed. One knot = 1.85 kilometer per hour.

Wind direction is determined with a **wind vane**. A wind vane is a pointer mounted on a shaft that is attached to a base marked with compass directions. The tail of the pointer has a larger surface area than the tip. Thus, the wind exerts more pressure on the tail than on the tip. This causes the tail to swing around so the tip is pointing in the direction from which the wind is blowing. Wind direction is given as the direction from which the wind blows.

Clouds and precipitation

Clouds and precipitation form when air is cooled below its dew point and water vapor condenses into tiny water droplets or ice crystals. In order for a gas to condense into a liquid, it must have a surface on which to condense. In the atmosphere, this surface is pro- vided by solid particles suspended in the air such as salt or ice crys- tals, dust, or smoke called *condensation nuclei*. Water vapor in the air condenses when it is cooled to the dew point temperature and there are condensation nuclei on which the condensation can form. If the air temperature is below the freezing point of water, water vapor may turn directly into ice crystals (called deposition).

How a cloud forms

In the troposphere, air temperature drops steadily with altitude, so as warm air rises it cools. When it has been cooled to its dew point, the water vapor in the air condenses. This condensation creates a **cloud** made of a great many tiny water droplets or ice crystals. We see a cloud by the light that reflects from these droplets or crystals. The base of the cloud shows the altitude at which the rising air reached its dew point temperature. Since this depends upon the amount of water vapor in the air and the starting temperature, clouds may form at many different altitudes.

Types of clouds

Clouds are important weather variables for three reasons: they are the source of precipitation, such as rain, snow, sleet, and hail; they reflect a lot of sunlight causing unequal heating of Earth's surface; and they are key indicators of overall weather conditions. Clouds are named by their altitude and vertical development.

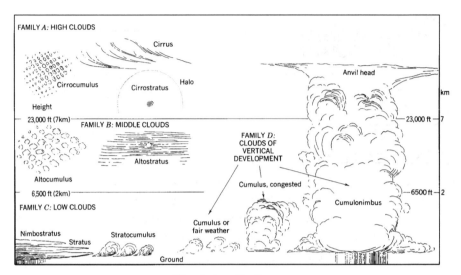

Cloud classification.

Fog, dew, and frost

Fog is a cloud whose base is at ground level. It forms when moist air at ground level is cooled below its dew point. When air condenses on a surface whose temperature is above freezing, droplets of water called *dew* form on the surface. If the temperature is below freezing, tiny ice crystals called *frost* will form instead of dew.

Precipitation

The water droplets or ice crystals in clouds are very tiny (~1/50 millimeter) and often remain suspended in the air. Thus, all clouds do not form precipitation. Precipitation occurs only when cloud droplets or ice crystals grow large and heavy enough to fall. **Precipitation** is any form of water that condenses in the atmosphere and falls to Earth's surface. The table on page 71 summarizes the different types of precipitation and how they form.

Types of precipitation.

Name	Description	Origin
Rain	Droplets of water up to 4 mm in diameter	Cloud droplets coalesce when they collide
Drizzle	Very fine droplets falling	Cloud droplets coalesce slowly and closely together when they collide
Sleet	Clear pellets of ice	Raindrops that freeze as they fall through layers of air at below-freezing temperatures
Glaze	Rain that forms a layer of ice on surfaces it touches	Supercooled raindrops that freeze as soon as they come into contact with below-freezing surfaces
Snow	Hexagonal crystals of ice or needlelike crystals at very low temperatures	Water vapor sublimes, forming ice crystals on condensation nuclei at temperatures below freezing
Hail	Balls of ice ranging in size from small pellets to as large as a softball with an internal structure of concentric layers of ice and snow	Again and again, hailstones are hurled up by updrafts in thunderstorms and then fall through layers of air that alternate above and below freezing—each cycle adds a layer to the hailstone—the more violent the updrafts, the larger and heavier the hailstone can become before falling

Measuring precipitation

Precipitation is measured with a rain gauge. A rain gauge is a simple instrument, usually a tube with markings on the outside. As rain falls, water collects in the tube. The amount of rain that has fallen is determined by reading the markings on the tube. In the United States, precipitation is measured in inches. If snow falls, it is melted and the level of the water is measured. Snow depth is measured directly using a ruler.

BRAIN TICKLERS   Set # 1

Match each of the weather variables with the weather instrument used to measure it.

Weather Variable	Weather Instrument
1. Air temperature	a. Anemometer
2. Air pressure	b. Psychrometer
3. Humidity	c. Thermometer
4. Wind speed	d. Wind vane
5. Wind direction	e. Rain gauge
6. Precipitation	f. Barometer

7. A northwest wind blows from the _____ toward the _____.

8. The closer the wet-bulb and dry-bulb temperatures, the _____ the relative humidity.

9. As the amount of water vapor in the air increases, the air pressure _____.

10. As air temperature rises, air pressure generally _____.

11. Which process most directly results in cloud formation?

    a. Condensation

    b. Transpiration

    c. Precipitation

    d. Radiation

Base your answers to questions 12 through 14 on the diagram on page 70 representing types of clouds and where they are found in Earth's atmosphere.

12. Identify *one* factor represented in the diagram that is used to classify the types of clouds.

13. Identify the water cycle process that forms clouds when moist air cools as it rises.

14. Explain why less sunlight reaches Earth's surface when cumulonimbus clouds are over a location than when cirrus clouds are over the same location.

(Answers are on page 101.)

Weather Patterns

When weather variables are plotted on a map, large-scale patterns can be seen. Tracking weather patterns over time reveals how and where the air and its weather are moving. This information can then be used to predict, or **forecast,** weather changes. A map that summarizes weather variables measured at many places at the same time is called a **synoptic weather map** (a synopsis is a summary).

The synoptic weather map

A synoptic weather map is made by measuring weather variables at thousands of weather stations around the world four times a day. These data are then used to create maps that reveal large-scale weather patterns. By looking at a series of synoptic weather maps, the development and movement of weather systems can be tracked and predictions can be made.

The station model

Synoptic weather maps use a symbol called a **station model** to show a summary of the weather conditions at a weather station. The chart below shows the typical symbols used on weather maps, including a typical station model, and explains what each of its elements represent.

Key to Weather Map Symbols

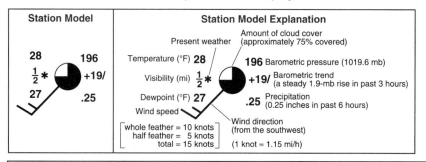

> ## PAINLESS TIP
>
> **Converting the three-digit air pressure value on a station model to atmospheric pressure in millibars**
>
> Air pressure rarely rises above 1050.0 mb or drops below 950.0 mb. The general rule is to place a 9 in front of the three digits on the station model if they are greater than 500, a 10 in front of the digits if they are less than 500, and a decimal between the last two digits. In the Station Model Explanation, the air pressure is listed as 196 (1019.6 mb). Because 196 is less than 500, a 10 is placed in front of the digits (10196) and a decimal is placed between the last two digits (1019.6 mb).

Station models plotted on a map summarize a wide range of weather data and can be used to create many different field maps. As you learned in Chapter 1, one way to represent field quantities on a two-dimensional field map is to use isolines. Isolines connect points of equal field values. For example, an air temperature field map would contain lines connecting points of equal temperature, or **isotherms**. An air pressure field map would contain lines connecting points of equal pressure, or **isobars**. Field maps on the following pages clearly show the weather patterns in the atmosphere.

(a) Synoptic weather map.

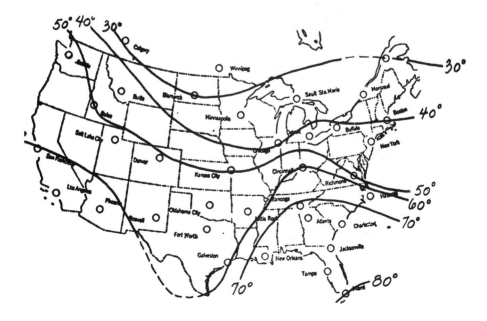

(b) Isotherms.

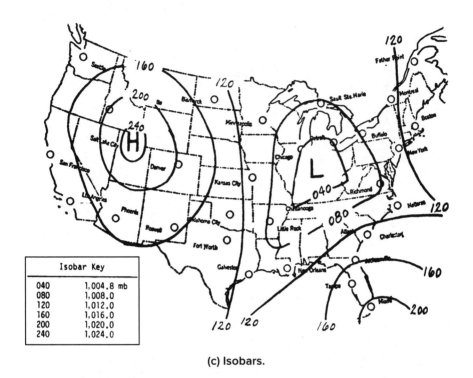

(c) Isobars.

As you can see in the synoptic weather map, there are distinct regions in the atmosphere that have similar conditions. For example, there is a region of high pressure, cooler temperatures, and clear skies centered near Salt Lake City, Utah. There is also a region of low pressure, warmer temperatures, and cloudy skies centered near Cincinnati, Ohio.

BRAIN TICKLERS   Set # 2

1. Which station model indicates a wind speed of 25 knots?

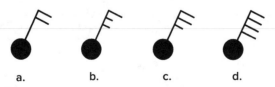

a.   b.   c.   d.

2. What is the air temperature shown on the station model below?

a. 8°F

b. 21°F

c. 50°F

d. 70°F

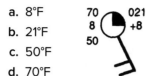

3. Two weather stations are located near each other. The air pressure at each station is changing so that the difference between the pressures is increasing. The wind speed between these two locations will probably

a. decrease.

b. increase.

c. decrease, then increase.

d. remain the same.

4. The U.S. weather map below shows weather data plotted for a December morning.

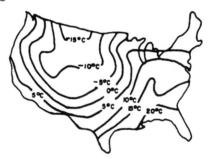

The isolines shown on the map most likely are

a. contour lines.

b. latitude lines.

c. isobars.

d. isotherms.

Base your answers to questions 5–8 on the chart below, which contains a summary of weather observations taken at noon over a six-day period.

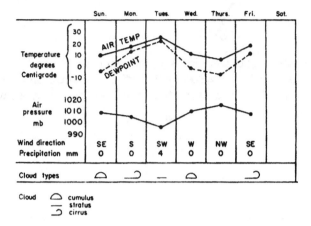

5. Between which two days did the direction of the wind shift 180°?

a. Sunday and Monday

b. Monday and Tuesday

c. Wednesday and Thursday

d. Thursday and Friday

6. On which of the following days was the relative humidity the lowest?

a. Sunday

b. Monday

c. Tuesday

d. Friday

7. On which day did the greatest chance of precipitation exist?

a. Monday

b. Tuesday

c. Wednesday

d. Thursday

8. Which weather symbol would most likely have been used to represent this region on Thursday?

26 Ⓛ 990   43 ◯ 001   32 ● 200   43 ◯ 150
15   33   -8   18

a.   b.   c.   d.

(Answers are on page 101.)

Air masses

Most weather forecasts are based upon the movements of large regions of air with fairly uniform characteristics, or **air masses**. When air rests on or moves slowly over a surface, it becomes like that surface. For example, air over the Gulf of Mexico in the summer rests on very warm water. The air is warmed by contact with the water and a lot of water vapor evaporates into the air. As a result, the air becomes warm and moist, just like the Gulf of Mexico; it becomes a warm, moist air mass. The longer the air lingers over a region, the larger the air mass becomes and the more like the surface beneath it the air mass becomes. The surface over which an air mass forms is called its **source region**. On weather maps, air masses are named for their source region as shown in the chart on page 80. Air masses are moved by global winds. Air masses that affect U.S. weather and their source regions are shown below. By examining a series of synoptic weather maps, the movements of air masses can be tracked and predictions about future movements can be made.

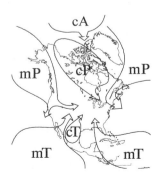

Air mass source regions.

	arctic A	polar P	tropical T
	Formed over extremely cold, ice-covered regions	Formed over regions at high latitudes where temperatures are relatively low	Formed over regions at low latitudes where temperatures are relatively high
maritime m Formed over water, moist		**mP**—cold, moist Formed over North Atlantic, North Pacific	**mT**—warm, moist Formed over Gulf of Mexico, middle Atlantic, Caribbean, Pacific south of California
continental c Formed over land, dry	**cA**—dry, frigid Formed north of Canada	**cP**—cold, dry Formed over northern and central Canada	**cT**—warm, dry Formed over southwestern United States in summer

Air mass names.

Weather fronts

At any given time, there are usually several air masses moving across the United States. They generally move from west to east driven by global winds. When different air masses meet, very little mixing of the air takes place. The whole surface along which the air masses meet is called the frontal surface; the line on the ground marking the boundary between the air masses is called a **front**. This boundary is called a front because it marks the leading edge, or *front* of an air mass that is pushing against another air mass. Fronts are places where weather changes rapidly and often have unsettled and rainy weather. The types of fronts and the symbols used to represent them are shown in the "diagrams on page 81."

Cold fronts

Cold fronts form along the leading edge of a cold air mass advancing against a warmer air mass. The cold air mass is denser than the warmer air ahead of it, so it pushes against it and under it like a wedge. This forces the warmer air upward rapidly and results in turbulence, rapid condensation forming heavy, vertically developed clouds, and heavy precipitation, or thunderstorms. The temperature drops sharply and the pressure rises rapidly when a cold front passes.

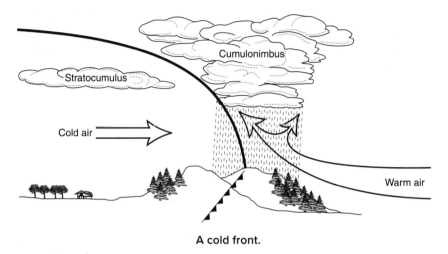

A cold front.

Warm fronts

A warm front forms where a warm air mass pushes up and over a cold air mass ahead of it. The warm air mass rides up and over the cold air because warm air is less dense than the cold air. As the warm air rides up over the cold air mass, it rises, expands, and cools. This causes condensation to occur over the wide, gently sloping boundary. The result is thickening, lowering clouds and widespread precipitation.

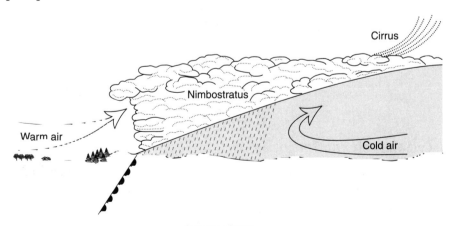

A warm front.

Stationary fronts

A stationary front forms between a warm air mass and a cold air mass when neither has enough force behind it to move the other. Stationary fronts slowly take on the shape and characteristics of a warm front as the denser cold air slowly slides beneath the warmer, less dense air. Stationary fronts have the widespread rain and cloudiness of a warm front. But the rain and clouds may linger for many days until another air mass comes along with enough impetus to get the stalled air masses moving.

Occluded fronts

Cold air is denser and exerts more pressure than warm air. Therefore, cold air masses and cold fronts tend to move faster than warm air masses and warm fronts. Therefore, a cold front will sometimes overtake a warm front. Then the warm air mass is trapped between two cold air masses and is lifted completely off the ground forming an **occluded front**. The lifting of the warm air mass causes a lot of condensation and precipitation resulting in widespread rain and thunderstorms. Depending on which cold air mass is denser, a cold front occlusion or a warm front occlusion may form.

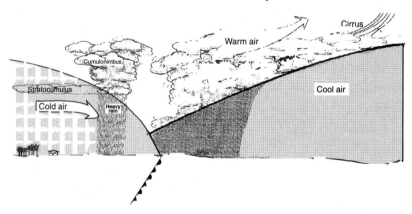

An occluded front.

Cyclones and anticyclones

Winds in the middle latitudes form weather systems called cyclones and anticyclones. These systems may stretch for several hundred to a thousand or more miles across and move from west to east. At the center of a **cyclone** the air pressure is low, and air rushing into it curves to the right in the Northern Hemisphere and toward the left in the Southern Hemisphere. As a result, winds in a cyclone blow in a counterclockwise spiral in the Northern Hemisphere and a clockwise spiral in the Southern Hemisphere.

An **anticyclone** is centered on a high-pressure region from which air moves outward. The Coriolis effect therefore causes winds in an anticyclone to blow outward in a clockwise spiral in the Northern Hemisphere and a counterclockwise spiral in the Southern Hemisphere. These spirals can be seen clearly in cloud formations photographed from Earth satellites.

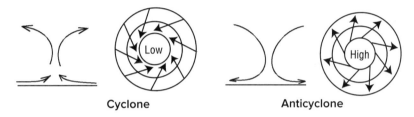

Cyclone   Anticyclone

Cyclone and anticyclone in the Northern Hemisphere.

In general, cyclones bring unstable weather with clouds, rain, strong winds, and sudden temperature changes. On the other hand, weather associated with anticyclones is pleasant, with clear skies and little wind.

How cyclones form

Cyclones in the United States form near our border with Canada. There lies the polar front, a zone where polar and tropical air masses meet. Polar winds blow from the northeast, whereas middle latitude winds blow from the southwest. Therefore, it is common for a bend to develop in this front with a bulge of warm air pushing into a cold air mass and vice versa. This produces a low-pressure region that moves eastward as a cyclone. The eastern side of the warm air wedge is a warm front, since warm air is pushing against cold air. The western side is a cold front, since cold air pushes against warm air. After a few days, the faster-moving cold front overtakes the warm front. The warm air is forced upward, cools, and soon afterward both it and the cyclone disappear.

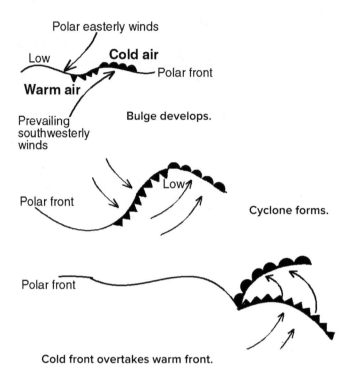

Cold front overtakes warm front.

BRAIN TICKLERS   Set # 3

1. The characteristics of an air mass depend mainly upon the

   a. rotation of Earth.

   b. cloud cover within the air mass.

   c. wind velocity within the air mass.

   d. surface over which the air mass was formed.

2. Which atmospheric gas has the greatest effect on the weather conditions associated with an air mass?

   a. Oxygen

   b. Nitrogen

   c. Methane

   d. Water vapor

3. What would an air mass that forms over the land in central Canada most likely be labeled?

    a. cP   **b.** cT   **c.** mT   **d.** mP

4. An air mass from the Gulf of Mexico moving north into the United States has a high relative humidity. What other characteristics will it probably have?

    a. Warm temperatures and low pressure

    b. Cool temperatures and low pressure

    c. Warm temperatures and high pressure

    d. Cool temperatures and high pressure

5. How does air circulate within a cyclone (low-pressure area) in the Northern Hemisphere?

    a. Counterclockwise and toward the center of the cyclone

    b. Counterclockwise and away from the center of the cyclone

    c. Clockwise and toward the center of the cyclone

    d. Clockwise and away from the center of the cyclone

    Base your answers to questions 6–9 on the diagram below. The diagram shows a cross section of weather systems over a part of North America that includes five weather stations: A, B, C, D, and E. The dashed lines represent frontal surfaces.

    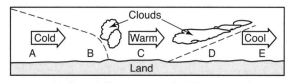

    Horizontal distance scale (in kilometers)

6. Which location is probably experiencing the warmest air temperature?

    a. A   **b.** B   **c.** C   **d.** D

7. Which symbol should be used on a weather map to represent the frontal surface between stations B and C?

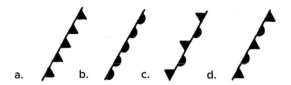

a.   b.   c.   d.

8. The frontal surface between stations C and D is most likely

   a. a warm front.   c. a stationary front.

   b. a cold front.   d. an occluded front.

9. The cold front is moving faster than the warm front. What usually happens to the warm air that is between the two frontal surfaces?

   a. The warm air is forced over both frontal surfaces.
   b. The warm air is forced under both frontal surfaces.
   c. The warm air is forced over the cold frontal surface and under the warm frontal surface.
   d. The warm air is forced under the cold frontal surface and over the warm frontal surface.

10. Why do clouds usually form at the leading edge of a cold air mass?

   a. Cold air contains more water vapor than warm air does.
   b. Cold air contains more dust particles than warm air does.
   c. Cold air flows over warm air, causing the warm air to descend and cool.
   d. Cold air flows under warm air, causing the warm air to rise and cool.

(Answers are on page 101.)

Weather Forecasting

Most weather forecasts are based on synoptic weather maps showing the movements of air masses and fronts. Synoptic weather maps are made four times a day: at midnight, 6 A.M., noon, and 6 P.M. By looking at a sequence of these maps, meteorologists keep track of the direction and speed at which the air masses and fronts are moving. They then project where they will move in the future. These are fairly accurate for large-scale, short-term (one to three days) forecasts. But local conditions can strongly affect weather. For example, the concrete and pavement of cities causes them to get hotter than the forests and fields of the surrounding countryside. This creates urban "heat islands" that can affect the path of a cyclone, the times when rain will begin and end, and the temperatures the area will experience.

Today, computers are used to collect, store, and transmit weather reports to meteorologists everywhere. They are used to plot weather maps showing patterns in all of the weather variables. They are also used to create complex computer models of the atmosphere based on the physics of the heat transfer and fluid behavior. These models are then used to predict changes in weather patterns. Because the atmosphere is constantly changing, forecasts become less reliable as they try to predict farther and farther into the future.

PROBABILISTIC FORECASTING

Have you ever heard someone say, "Weather forecasters are the only people who get paid to be wrong every day!"? Very funny. But what most people don't realize is just how much complicated science and math is involved in making good forecasts. A forecast is just that, a prediction, an educated guess. Being wrong is just a part of a weather forecaster's job.

Most weather forecasting is probabilistic. Mathematical equations are used to predict the probability that the atmosphere will behave in a certain way based on the past behavior of the atmosphere.

Basically, the system works something like this: Suppose you have math first period on Mondays, Tuesdays, and Wednesdays, and science first period on Thursdays. At first glance the first period pattern looks simple: MMMS. However, there would be exceptions. There might be a holiday on a Monday, so the sequence would be MMS. Or Thursday might be a holiday, so the sequence would be MMMMMMS. However, such deviations in the schedule would probably be the exception rather than the rule.

Let's assume that in most previous school years, on average there was a deviation every four weeks. Probabilistic forecasting would then say that for the next week there was a three in four, or 75 percent, chance that three maths would be followed by a science.

Now let's apply this reasoning to weather. Suppose we observed that on average, three out of the past four times the wind shifted to blow from the northeast, we had rain within 6 hours. Then, if the wind shifted to the northeast, probabilistic forecasting would say that there was a 75 percent chance of rain in the next 6 hours.

Doing probabilistic forecasting involves three basic steps: maintaining detailed historical records, identifying all previous occurrences of a pattern, and then calculating the probabilities. However, there is no way to anticipate unusual behavior, nor is past behavior a guarantee of future behavior. So, forecast errors are to be expected and are just part of the job for a meteorologist.

Weather hazards

Hazardous weather, such as tornadoes, severe thunderstorms, hurricanes, and winter storms, claims many lives and causes much property damage each year in the United States. They all involve low pressure, clouds, precipitation, and strong winds, but each has its own characteristics and dangers.

Thunderstorms

Thunderstorms are likely to occur wherever and whenever there is strong heating of Earth's surface. This causes warm air to rise rapidly forming a cloud that grows larger and larger as more and more warm, moist air is carried upward. The strong updrafts in the rising air support water droplets and ice crystals in the clouds so that they grow in size. When the updrafts cannot support the moisture any more it falls as rain, or even hail. The falling rain sets up downdrafts that cause internal friction with updrafts. The internal friction builds up static electric charges that may discharge as lightning.

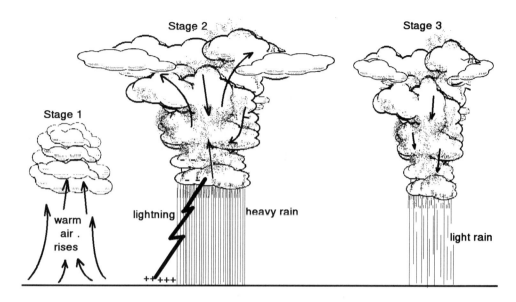

Formation of a thunderstorm.

Hurricanes

Hurricanes are huge cyclonic storms that form over oceans near the equator during the summer months, when the ocean surface is the warmest. Heat and water vapor enter the air from the warm ocean. Both heat and water vapor decrease the air's density and lower the air pressure. The combination of the two causes air pressure to get very low. This causes many strong convection cells to form, and with them, many thunderstorms. If the thunderstorms merge, they can form a huge convection cell. As strong winds carry heat and moisture toward the low-pressure center, the convection cell gets larger and stronger. When winds reach 119 kilometers per hour, the storm is called a hurricane. Fully formed hurricanes are huge cyclones, often exceeding 500 kilometers in diameter. As these storms spin, cool air above the storm sinks into the center of the cyclone. This air is sinking, not moving horizontally, so winds are calm. The sinking air is warming by compression, so there is little condensation and skies are clear. This calm, clear at the center of the hurricane, is called the **eye** of the hurricane.

Anything that cuts off its supply of heat or moisture will weaken a hurricane, so hurricanes lose strength as they move over land or cool water. However, even after moving over land, hurricanes can last for many days and cause much damage due to high winds and flooding.

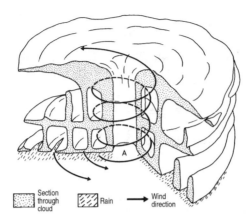

Section through cloud   Rain   → Wind direction

Cross-section of a hurricane.

Tornadoes

Tornadoes are small (most are less than 100 meters in diameter), brief (most last only a few minutes) disturbances that usually form over land from intense thunderstorms. Tornadoes usually form late in the day, when Earth's surface is the warmest. When heating is very intense, warm air rises in strong convection currents. The upward movement of the air causes a sharp decrease in pressure. Air rushes into the low-pressure region from the sides and is given a spin by the Coriolis effect. Wind speed gets very high near the center of the updraft due to the big difference in air pressure. The rapidly moving air decreases the pressure even more, which further feeds the updraft. The whole process spirals upward in intensity and a funnel forms that eventually touches the ground. Wind speeds near the center of a tornado may reach speeds of 500 kilometers per hour or more.

Winter storms

In many areas of the country, winter cyclones bring heavy snowfall and very cold temperatures. Heavy snow can block roads and cause power lines to fall down. The cold temperatures can be dangerous if a person is not properly dressed. Winter storms include ice storms and blizzards. In an **ice storm**, rain freezes when it hits the ground, creating a coating of ice on roads and walkways. Rain that turns to ice pellets before reaching the ground is called sleet. Sleet also causes roads to freeze and become slippery. In a **blizzard**, heavy snow and strong winds produce a blinding snow, near zero visibility, deep drifts, and life-threatening wind chill. A major winter storm can be lethal; if possible, stay indoors and do not travel.

Emergency preparedness

Hazardous weather happens; there's not much we can do about that. But advance planning and quick response can help you survive hazardous weather. The National Weather Service issues several levels of alerts about severe weather: "watches," "warnings," and "advisories." A **watch** means that hazardous weather conditions are likely to develop in your area, but its occurrence, location, or timing is still uncertain. If a watch is issued for your area, you should stay alert to the weather by listening to the radio or television and be prepared to take shelter. A **warning** means that hazardous weather has been sighted or shown on radar in your area. If a warning is issued, the danger is very serious and everyone should go to a safe place. Once there, listen to a battery-operated radio or television for further instructions. Wait for an official "all clear" before resuming normal activities. An **advisory** is issued when hazardous weather is imminent or occurring. However, advisories are issued for less serious conditions than warnings—conditions that may cause significant inconvenience and, if caution is not exercised, could threaten life or property. The Federal Emergency Management Agency has developed severe weather fact sheets that describe specific steps to be taken when watches, warnings, and advisories are issued.

BRAIN TICKLERS   Set # 4

1. Present-day weather predictions are based mainly upon the movements of _____ and _____.

2. A hurricane releases more energy in one day than is consumed by people worldwide in one year. Where does all of this energy come from?

3. In one or more sentences, explain the difference between a tornado "watch" and a tornado "warning."

4. Which diagram best represents the air circulation as seen from above in a low-pressure center hurricane in the Northern Hemisphere?

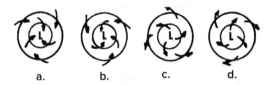

a.   b.   c.   d.

5. Most thunderstorms form because intense heating of Earth's surface causes air to

   a. rise rapidly forming a cloud that grows larger and larger.
   b. become denser and sink rapidly to the ground.
   c. become so hot it forms lightning.
   d. rise more slowly so raindrops have more time to form.

(Answers are on page 102.)

Climate

Climate refers to the general weather conditions in a region over a long period of time. One cannot determine the climate of a region by looking at the weather over a single year. In 1988, the Midwest suffered a near rainless summer. Yet, in the summer of 1993, many of these same places were drenched by torrential rains. Many people wondered, are we seeing Earth's global climate change? One or two extreme summers can't answer that question. Our picture of climate develops slowly as we watch dozens of seasons pass. Some winters are warmer than others, some summers drier, some falls colder. We can only get a sense of the shifting patterns of climate by comparing measurements taken over many years and decades.

The main elements of climate
Temperature

Climate is the result of the interplay of a number of factors. One of the most important is energy. Earth's main source of energy is sunlight, which warms the land, which, in turn, heats the atmosphere. A key way of tracking energy flow in a region is by monitoring its temperature. Therefore, scientists who study climate keep track of air temperatures over land and sea, air temperatures at various altitudes, and ocean water temperatures around the globe.

Moisture

The other major climate factor is water. Water is important to all living things. To a large extent, it controls the type of plant and animal life that can live in a region. It is also important in weathering and is the main agent of erosion and deposition on Earth. Some of the measurements used to track water and climate changes are precipitation, the amount of water vapor in the air, the amount of snow and ice cover on land, and the extent of sea ice.

Some factors affecting climate

The main elements of a region's climate are its temperature and precipitation patterns. Temperature and precipitation patterns are controlled by various factors, such as latitude, nearness to a large body of water, elevation, mountain ranges, and vegetation.

Latitude

Latitude is the main factor affecting a region's temperature patterns. Latitude determines both the angle at which sunlight strikes Earth's surface *and* the length of the daylight period during which sunlight can warm the region. Differences in these two things result in three main temperature zones on Earth: the always cold **frigid zone** near the poles, the always hot **torrid zone** near the equator, and the seasonally changing **temperate zones** in between.

Large bodies of water

Land surfaces heat up and cool off faster than water surfaces. Therefore, air temperatures are usually colder in the winter and warmer in the summer over land masses than they are over oceans at the same latitude. Large bodies of water moderate the temperatures of nearby land by warming it in the winter and cooling it in the summer. Therefore, cities near a large body of water tend to have warmer winters and cooler summers than cities far inland.

Elevation

In the troposphere, there is a gradual, but steady, decrease in temperature with elevation. Temperature decreases about 1°C for every 100-meter rise in elevation. This explains why high mountains may have tropical jungles at their base, but permanent ice and snow at their peaks. In general, a city at a high elevation will have a cooler climate than one at a low elevation.

Mountain ranges

Mountain ranges serve as barriers to outbreaks of cold air. In this way, the Alps protect the Mediterranean coast and the Himalayas protect India's lowlands from cold, polar air. The side of a mountain range facing the prevailing winds tends to have a cool, moist climate, whereas the other side of the mountain range has a warmer, drier climate. This happens because air that is blown over mountain ranges is forced to rise and cool. This causes condensation that forms clouds and precipitation. By the time the air reaches the top of the mountains it has lost much of its moisture. When the air descends on the other side of the mountains it is warmed by compression. The air is warmer and drier and precipitation is less likely. This explains why Tillamook, a city west of the Cascade Mountains in Oregon, has a cool, wet climate, whereas Bend, Oregon, on the other side of the Cascades, has a warmer, drier climate.

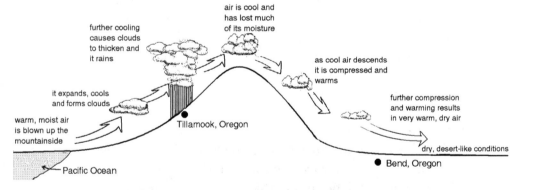

Tillamook and Bend have different climates because they are on opposite sides of the Cascade Mountains.

DEFORESTATION

Wood is a valuable natural resource. It is a fuel, a building material, and a raw material for other products such as paper. The demand for wood has led to deforestation. Deforestation has had far reaching effects on the environment.

Trees absorb CO_2 and emit O_2 as they carry out photosynthesis. Photosynthesis ties up CO_2 in the form of sugars, starches, and cellulose. Deforestation decreases the amount of carbon dioxide removed from the atmosphere by trees, causing a net increase in carbon dioxide in the atmosphere. Deforested land is also susceptible to erosion and loses much of its fertile topsoil during rainstorms. The topsoil ends up in streams and other bodies of water where it clogs channels and kills aquatic organisms. When land is deforested, many organisms are left without a habitat and eventually die. Deforested land also heats up and cools off faster than forested land, causing a change in both local weather patterns and climate.

Vegetation

Vegetation affects the climate by influencing the processes of transpiration and surface runoff that, in turn, influences rainfall. Vegetation also influences temperature. Vegetation reflects less insolation back into space than bare surfaces, which tends to warm the climate. But this effect is small compared to the cooling effect when vegetation absorbs carbon dioxide from the air and thereby decreases the greenhouse effect. Thus, deforesting the land tends to warm the climate.

Types of climates

Climates are classified by moisture, temperature, and vegetation patterns. There are several systems in use today. The terms often used to describe different patterns of moisture, temperature, and vegetation are shown in the table on page 96 and the map shows the major climates of the world.

Table of some descriptive terms for climate patterns.

Moisture	Temperature	Vegetation
Arid	Polar	Desert
Semiarid	Subpolar	Grassland, steppe, taiga
Subhumid	Subtropical	Deciduous forest Coniferous forest
Humid	Tropical	Rain forest

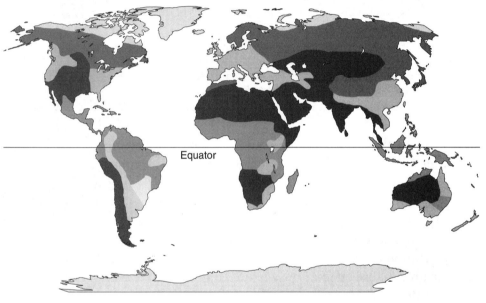

Equator

Key

Polar climate – very cold year-round, ground always frozen; rainfall <10 in/year

Warm temperature climate – hot summers, cold winters; rainfall 30-60 in/year

Tropical rainy climate – hot all year, rainfall up to 80 in/year

Cool temperature climate – warm summers, cold winters, rainfall <20 in/year

Dry and desert climate – hot summers, warm or cool winters; rainfall about 10 in/year

World climate zones.

RE-DRAWING THE MAP

Climate regions worldwide are changing as you read this! Human activities, such as the release of greenhouse gases from the burning of fossil fuels and the loss of forests that would otherwise store CO_2, are driving global warming and climate change. For example, the dry edges of the tropics are expanding toward the poles, while the wettest area near the equator is shrinking. In the U.S., plant hardiness zones that track the types of plants that will do well in an area are shifting northward. Deserts are growing in size. The dry region in the western U.S. has expanded 140 miles east since 1980. In 30 years, "tornado alley" in the U.S. has shifted 500 miles east. Areas of ice and permafrost are shrinking. Plants, animals, and diseases are on the move as the climate that suits them moves. Some species are becoming extinct as they lose habitat. And, as the climate moves, so are the lines on this map.

BRAIN TICKLERS   Set # 5

1. Place a "W" next to statements that describe weather and a "C" next to statements that describe climate.

   a. The air temperature in Dallas, Texas, was 82°F at 3:00 P.M. today.

   b. On average, Chicago, Illinois, has 17 days with maximum temperatures in excess of 90°F each year.

   c. During January 2010, high temperatures in Miami, Florida, were 5°F below average and rainfall was 2–3 inches above average.

   d. Skies in Los Angeles, California, were clear this morning but became cloudy this afternoon.

   e. Yesterday's snowstorm dropped 32 inches of snow on Denver, Colorado.

   f. Average rainfall is highest in Dallas, Texas, during May and October.

2. As elevation increases, average annual temperature
   _____.

3. Which graph best represents the general relationship between latitude and average surface temperature?

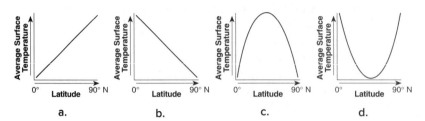

a.   b.   c.   d.

4. Large bodies of water have a moderating effect on climate primarily because

   a. water gains heat more rapidly than land does.

   b. water surfaces are flatter than land surfaces.

   c. water temperatures are always lower than land temperatures.

   d. water temperatures change more slowly than land temperatures.

Base your answers to questions 5 and 6 on the diagram below, which shows air movement over a mountain range. The arrows indicate the direction of airflow. Points 1 through 3 represent locations on Earth's surface.

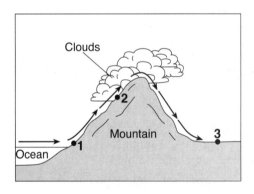

5. Compared to the climate at location 1, the climate at location 3 is

   a. cooler and drier.   **c.** warmer and drier.

   b. cooler and wetter.   **d.** warmer and wetter.

6. Cloud formation at location 2 is the direct result of air that is rising,

   a. expanding, and cooling.

   b. compressing, and cooling.

   c. expanding, and warming.

   d. compressing, and warming.

7. The chart below shows the range of total annual precipitation for areas with different kinds of climates.

Area	Annual Precipitation (cm)
Tundra	10–40
Grassland	10–60
Desert	0–20

Which graph best represents the data in the table?

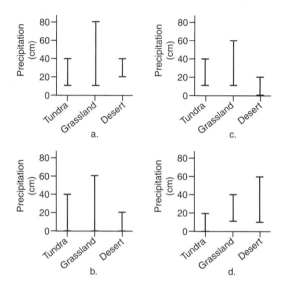

(Answers are on page 102.)

Wrapping up

- Weather is the condition of the atmosphere at a particular place and time.
- Weather is described by measurements of weather variables (characteristics of the atmosphere) such as temperature, humidity, precipitation, air pressure, and wind.
- A map that summarizes weather variables measured at many places at the same time is called a synoptic weather map.
- An air mass is a large region of air with relatively uniform characteristics. A front marks the boundary between two different air masses.

- Most weather forecasts are based on the movements of air masses and fronts.

- Climate refers to the long-term weather conditions in a region. Climate is determined by averaging measurements of weather factors over a long time period (years/decades).

- The main elements of a region's climate are its temperature and precipitation patterns. Temperature and precipitation patterns are controlled by various factors, such as latitude, nearness to a large body of water, elevation, mountain ranges, and vegetation.

Brain Ticklers—The Answers
Set # 1, page 72

1. c	6. e
2. f	7. northwest; southeast
3. b	8. higher
4. a	9. decreases
5. d	10. decreases
	11. a

12. altitude/elevation/height, *or* cloud shape, *or* temperature

13. condensation *or* a change from gas to liquid.

14. Cumulonimbus clouds are thicker. *or* Cirrus clouds are thinner. *or* Cumulonimbus clouds reflect, absorb, and/or block more sunlight.

Set # 2, page 76

1. c   2. d   3. b   4. d   5. d   6. a   7. b   8. d

Set # 3, page 84

1. d	6. c
2. d	7. a
3. a	8. a
4. a	9. a
5. a	10. d

Set # 4, page 91

1. air masses; fronts

2. The Sun

3. A watch means hazardous weather conditions are likely to form in your area. A warning means hazardous weather has been sighted or shown on radar in your area.

4. b

5. a

Set # 5, page 97

1. a. W; b. C; c. C; d. W; e. W; f. C

2. decreases

3. b   4. d   5. c   6. a   7. c

Chapter 4

The Hydrosphere

Earth is often called "The Water Planet" because water covers about 70 percent of its surface. All of the water blanketing Earth's surface is called the hydrosphere. The hydrosphere includes all of Earth's oceans, lakes, streams, underground water, and ice. A tiny fraction exists in the atmosphere as water vapor. Although the hydrosphere is made up of water, it is not all plain liquid water. About 97 percent of the hydrosphere is seawater and 3 percent is freshwater. Of the 3 percent freshwater, 2 percent is frozen or unavailable in some other way. That doesn't leave much water for the almost 8 billion people on the planet to use.

Distribution of Earth's Water

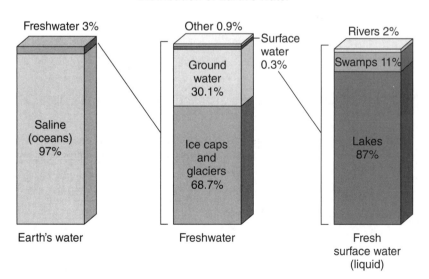

How the Hydrosphere Formed

The gases released from Earth and carried in by comets that formed our current atmosphere were rich in water. (A typical comet contains about 10^{15} kilograms of water—enough water to entirely fill the Great Lakes!) Because Earth was very hot, the water in Earth's newly developed atmosphere was water vapor. Then, as Earth cooled, the water vapor in the atmosphere condensed and fell as torrential rains. At first, Earth's surface was still so hot that the water quickly vaporized and returned to the atmosphere. But as Earth cooled further, the water remained on its surface longer and longer as a liquid. Eventually, Earth's oceans and lakes filled with water that only returned to the atmosphere by evaporation due to the Sun, as it does today. When the atmosphere had cooled enough for some of the water vapor in the atmosphere to condense as snow or ice crystals and fall to Earth's surface, glaciers and ice caps formed. As water continues to work its way to the surface or is carried in by comets, the hydrosphere may gradually increase in volume until, in a few billion years, it covers Earth's entire surface.

BRAIN TICKLERS   Set # 1

1. All of the water blanketing Earth's surface is called the

   _____.

2. Outgassing and _____ added large amounts of water vapor to Earth's atmosphere.

3. Earth's hydrosphere formed when the water vapor in Earth's _____ condensed and fell as rain.

4. Place the following parts of the hydrosphere in order from greatest to smallest percentage of the hydrosphere: fresh-water lakes, oceans, ice caps and glaciers, and rivers.

   Greatest _____; _____; _____; _____ Smallest

(Answers are on page 127.)

The Ocean

When the hydrosphere condensed from the atmosphere, the water ran downhill and collected in the lowest lying areas of Earth's surface. Over time, these depressions filled to overflowing with water, became interconnected, and created a worldwide ocean. The Atlantic, Pacific, Indian, and other "oceans" are only subdivisions of this continuous worldwide ocean that makes up about 97 percent of the hydrosphere.

The topography of the ocean floors

Have you ever wondered what Earth would look like if you removed the water covering its surface? Using many different kinds of technologies, scientists have mapped the bottoms of the oceans. The figure below shows what they have found beneath the oceans.

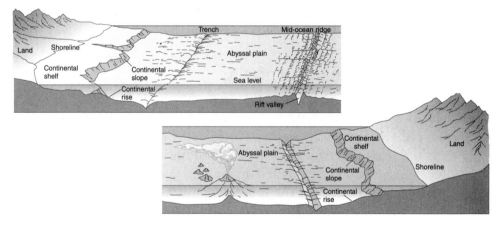

Topography of the ocean floors.

As you can see, the ocean floor near the shores is different from the floor near the center of the ocean. Most continents are fringed by a shallow region called the **continental shelf** that slopes gently toward the ocean basins. The continental shelves cover a large area, almost one-sixth of Earth's surface! Though fairly smooth, they also contain many small hills, valleys, and depressions. The edge of the continental shelf is marked by a steeply sloped region called the **continental slope** that ends in a gently sloping region called the **continental rise**. The continental rise is made up of small bits of rock along with plant and animal remains that were washed down from the continents and continental slopes.

Beyond the continental rise lies the **abyssal plain**, a wide, flat area of deep ocean. The abyssal plain is flat because any low or high spots have been filled in or covered by sediment washed into the deep ocean. The most striking feature of the ocean floors are the **mid-ocean ridges**, or undersea mountain chains running through the centers of the oceans. In some places the peaks of these ridges break the surface and form islands such as Iceland and the Azores. The mid-ocean ridges form a continuous feature that can be traced for more than 50,000 kilometers through the Atlantic, Pacific, Indian, and Arctic oceans. Many mid-ocean ridges are split by a deep **rift valley**. Rift valleys can plunge more than 2,000 meters below the surrounding peaks.

The deepest places in the oceans are called **trenches**. Trenches are deep, V-shaped valleys with narrow floors. They are usually no more than a few kilometers wide and more than 6 kilometers deep. The deepest trench is the Marianas Trench, which plunges to a depth of 11,022 meters. (See Brain Tickler on page 107.)

The composition of the oceans

The water that originally condensed from the atmosphere and fell to Earth's surface was fresh water. If you have ever taken a swim in the ocean and accidentally swallowed a little water, you know that the water in the oceans is no longer fresh, but quite salty. How did the water in the oceans become salty? Every time it rains, water running downhill dissolves some of the substances over which it flows. For billions of years, streams have carried these dissolved substances into the oceans. Every time water in the oceans evaporates, the dissolved substances are left behind. Over time, these dissolved substances have steadily built up in the waters of the oceans, creating the salty seawater we have today. The total amount of dissolved substances in seawater is called **salinity**. Salinity is measured in parts per thousand (0/00) by weight in 1 kilogram of seawater. The average salinity of seawater is about 35 parts per thousand.

Seawater is not just a simple mixture of salt and water. It is a complex mixture of dissolved inorganic and organic matter, dissolved gases, and particulates.

BRAIN TICKLERS   Set # 2

Write the letter of the term that best matches the definition.

1. A wide, flat area of deep ocean
2. Shallow zone that fringes the continents
3. Gently sloping zone at the base of the continental slope
4. Undersea mountain range running through the centers of the oceans
5. Deep, V-shaped valley with a narrow floor that forms the deepest places in the ocean

a. Continental rise
b. Continental shelf
c. Trench
d. Abyssal plain
e. Mid-ocean ridge

(Answers are on page 127.)

PAINLESS TIP

Organic—comes from living things. Remember that organs are found in living things.

Inorganic—not organic; not from living things.

- Inorganic matter comes mainly from minerals that dissolve in water. Most of the dissolved substances in seawater are inorganic compounds called salts.

- Most organic matter consists of wastes excreted by living things and the decaying bodies of dead organisms.

- Gases usually dissolve into seawater from the atmosphere. The major gases found dissolved in seawater are nitrogen, oxygen, and carbon dioxide. However, oxygen also enters seawater from plants living in the upper layers of the ocean (mainly phytoplankton) during photosynthesis.

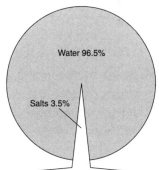

Water 96.5%	
Salts 3.5%	
Sodium chloride:	27.2 0/00
Magnesium chloride:	3.8 0/00
Magnesium sulfate:	1.7 0/00
Calcium sulfate:	1.3 0/00
Potassium sulfate:	0.9 0/00
Calcium carbonate:	0.1 0/00
Magnesium bromide:	0.1 0/00

The salinity of seawater. The symbol 0/00 means parts per thousand.

- Particulates are solid particles that are not dissolved in seawater, but are suspended in or settling through it. Most particulates are fine pieces of minerals and decaying organic material.

BRAIN TICKLERS   Set # 3

1. Give three types of dissolved substances that can be found in seawater.

   _____; _____; _____.

2. The number of grams of dissolved substances in 1 kilogram of seawater is called the seawater's _____.

3. Seawater has an average salinity of about _____ parts per thousand.

4. Over time, the salinity of the oceans has

   a. increased.   b. decreased.   c. remained the same.

5. A lake without any outlets may increase in salinity if the fresh water flowing into the lake is

   a. less than or equal to the amount of water lost by evaporation.

   b. greater than the amount of water lost by evaporation.

   c. not carrying any dissolved substances into the lake.

(Answers are on page 127.)

Circulation in the oceans

Like the atmosphere, the oceans are a fluid. And like the atmosphere, density differences in the oceans create convection currents. Density differences in the ocean are mainly due to differences in the temperature and salinity of the ocean waters.

If you swim in the ocean, you know that the water is warmest in the summer. During the summer season, sunlight is more direct and there are more hours of daylight. However, all year long the Sun's rays are most direct near the equator and least direct near the poles. Therefore, the surface waters of the oceans are warmest near the equator and coldest near the poles.

The salts that dissolve in seawater are denser than the water. Therefore, a mixture of water and salts is denser than water alone. Seawater of high salinity tends to sink in seawater of lower salinity at the same temperature because it is denser. What causes differences in salinity? Evaporation and freezing remove water from seawater but leave behind the dissolved substances causing an increase in salinity. Runoff from land and melting ice add freshwater to the seawater, thereby causing a decrease in salinity.

PAINLESS TIP

Salts are denser than water. Think of pure water as a pail filled with plastic marbles. Then imagine the same pail filled with a mixture of lead marbles (salts) and plastic marbles (water). The pail of mixed marbles would have a greater mass in the same volume and therefore a greater density. And, the higher the ratio of lead marbles to plastic marbles, the higher the density. The higher the salinity, the denser the seawater.

Convection currents

Differences in the temperature and density of water near the poles and equator creates convection currents in the ocean. Dense, cold water formed near the poles sinks and slowly flows along the ocean floors toward the equator. As the cold water moves toward the equator, it displaces warmer water upward causing *upwelling*. Upwelling brings the cold water's nutrients and oxygen to the surface, supporting a rich growth of plants and animals. It also causes the warmer water to spread outward away from the equator when it reaches the surface. The result is a vertical convection cell that transfers heat from the equator toward the poles.

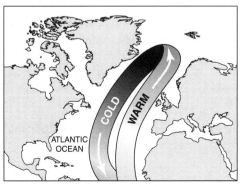

Convection currents in the oceans.

Wind-driven currents

Surface waters are not only set in motion by convection cells within the ocean. Winds exert a frictional force on the ocean surface and produce wind-driven currents. These currents roughly follow the pattern of global surface winds. For example, trade winds blowing from the east push surface waters to the west forming westward-moving currents near the equator (0°–30° latitude). In the Atlantic and Pacific Oceans, these westward currents are obstructed by land masses and are deflected north and south. These deflected currents, which run north and south along the western boundary of the oceans, are among the largest and strongest currents in the ocean. The Gulf Stream, a western boundary current in the Atlantic Ocean, runs north along the eastern coast of the United States and transports more than 100 times the output of all the rivers in the world.

When the western boundary currents reach the mid-latitudes (30°–60°), the prevailing winds, which blow from the west, push them back eastward across the ocean. The result is a large circular pattern of motion called a **gyre**. The major ocean currents shown here help distribute heat energy across Earth's surface and influence the weather and climate of nearby land masses.

THE GREAT PACIFIC GARBAGE PATCH

The circular motion of the North Pacific Gyre has created one of the largest patches of floating garbage on Earth. This huge swirling pile of trash, estimated at 1.6 million square kilometers in size, is more than 9 times the size of the entire state of New York! While many types of trash end up in the ocean, the majority is plastics. Most plastics are not biodegradable. That is, they do not decompose naturally. Instead, they simply break down into smaller and smaller pieces called microplastics.

Microplastics make up the bulk of the trash patch and make the water appear like a cloudy soup. Mixed in or floating on top are larger bits of debris like lost or abandoned fishing gear, plastic bags, plastic bottles and caps, Styrofoam cups, straws, wrappers, and even odd items like shoes or Lego blocks.

All of the plastic in the oceans is the result of human activities. It damages or destroys natural habitats and endangers marine life.

Fish, sea turtles, seabirds, and marine mammals can get entangled in or eat plastic debris. This can cause suffocation, drowning, starvation, and damage to internal organs.

Microplastics also block light needed by algae and plankton—the basis of the marine food web. Less light means less algae and plankton, which means less food for animals that feed on them, and so on up the food chain.

What is the best way to clean up the Great Pacific Garbage Patch? Most scientists and researchers agree that limiting or eliminating our use of disposable plastics and switching to biodegradable or reusable materials is the best option. We can all make a difference by the choices we make when buying.

Surface Ocean Currents

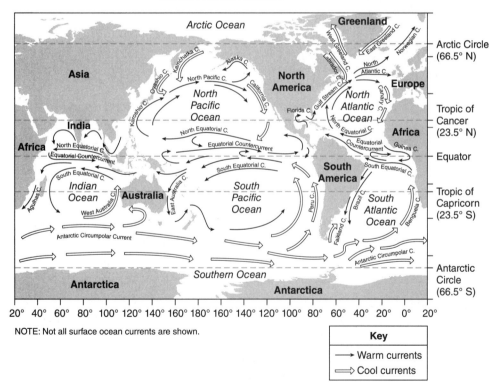

NOTE: Not all surface ocean currents are shown.

Key	
→	Warm currents
⇒	Cool currents

OCEANS AND CLIMATE

Ocean currents control climates by transferring heat from the equator toward the poles, thereby cooling the equatorial regions, warming the polar regions, and influencing the temperature of adjacent landmasses these currents pass. The major surface ocean currents (see page 111) follow the prevailing winds that blow out of the subtropical high-pressure belts, giving rise to a clockwise flow in the Northern Hemisphere and a counterclockwise flow in the Southern Hemisphere.

Wherever the ocean currents flow toward the poles, they carry warm water *from* the equator. Wherever they flow *toward* the equator, they carry cold water from the polar regions. Ocean currents such as the Gulf Stream and the North Atlantic Current carry warm water northward and eastward. The warm air over these ocean currents moderates the climate of the eastern coast of the United States, the Azores, and western Europe. The cold air over the icy waters of the Peru Current modifies the climate all along the western coast of South America, causing it to be much cooler and dryer than places with the same latitudes on the eastern coast of South America, where the Brazil Current and South Equatorial Current bring warm, moist air with their waters.

BRAIN TICKLERS   Set # 4

Circle the words in questions 1–3 that best complete the statement.

1. Deep ocean currents are formed when (cold, hot) water near the poles sinks to the ocean floor and flows (toward, away from) the equator.

2. Upwelling is caused by cold water flowing along the ocean (bottom, surface) toward the equator displacing (warmer, cooler) water upward.

3. The Peru Current along the west coast of South America and the Canaries Current along the west coast of Africa are both (warm, cool) currents that flow (toward, away from) the equator.

4. Which interaction between the atmosphere and the hydrosphere causes most surface ocean currents?

   a. The cooling of rising air above the ocean surface

   b. Evaporation of water from the ocean surface

   c. Friction between global winds and the ocean surface

(Answers are on page 127.)

Fresh Water

Less than 3 percent of Earth's water is fresh water. Of that, only about one-fifth is not frozen. This small percentage makes up the world's total available supply of fresh water. A typical person needs about 2–3 liters of fresh water each day to survive. However, people tend to use much more than that each day. An average person in the United States uses as much as 100 gallons (387 liters) of fresh water each day. On a global scale, the amount of fresh water humans use is staggering.

Why, then, has Earth's fresh water supply not run out? Earth's fresh water supply has not run out because it is continually being renewed. Every day, fresh water is produced as water evaporates, condenses, and falls back to Earth's surface as precipitation. When precipitation falls on land, a number of things may happen to it. It may evaporate, seep into the ground, or run off. Some precipitation may stay on the surface for a while as ice or snow. But when the ice or snow melts, it also either evaporates, seeps into the ground, or runs off.

Groundwater

Have you ever spilled a drink at the beach? What happens to the drink? It seeps right into the sand! After a storm, rainwater also seeps into the ground. Water that seeps into the ground becomes part of Earth's underground water supply, or **groundwater**.

Infiltration

The process by which water enters the ground is called **infiltration**. Infiltration occurs because Earth's surface is not completely solid. Much of Earth's surface is made up of loose particles of rock and soil. Between these particles there are open spaces called **pores**. Pores are usually filled with air unless water or some other substance has forced the air out. Precipitation and meltwater infiltrate Earth's surface by entering these pores and forcing the air out.

Porosity and permeability

The amount of water that can infiltrate depends on how much pore space is available for water to enter. It also depends on how easily water can enter and flow through the ground.

The amount of open space in a substance is its **porosity**. Porosity is expressed as a percent of a sample's total volume. Suppose that half the volume of a sample of sand is open space. Then that sample has a porosity of 50 percent. Porosity can be calculated by measuring the amount of water it takes to fill up the pores in a substance and then setting up a ratio of the volume of water to the total volume of the sample being tested.

For example, suppose a cup contains 100 milliliters of sand. When 35 milliliters of water have been poured in, the water is just even with the surface of the sand. This means that there were 35 milliliters of pore space in the sand. The porosity equals $35/100 \times 100$ or 35 percent.

The substances that make up Earth's surface vary greatly in porosity. Loose rock fragments like sand and gravel usually have the highest porosity. But even rocks that look solid may have some open spaces.

The porosity of a substance is influenced by a combination of many factors. These factors include the shape of its grains, how tightly together its grains are packed, and whether or not they are sorted. The diagrams below show how each of these factors affect the porosity of a substance.

 CAUTION—Major Mistake Territory!

A common mistake is to think that substances made of big particles are more porous than those made of small ones. This is because the pore spaces between large particles are clearly bigger than those between small particles. However, if you count the *total number of pore spaces* in a sample, you find that small particles have many more pore spaces than large particles. The volume of a small number of large pores and a large number of small pores turns out to be about the same. All other factors being equal, the porosity of large and small particles is about the same. Therefore, particle size does not affect porosity.

Permeability is the rate at which water can move through a substance. A substance through which water passes rapidly is said to be **permeable**; a substance through which water cannot flow is **impermeable**. The permeability of a substance depends on two factors: the size of its pores and the degree to which they are inter-connected. See the figure below. Small pores constrict the flow of water, and if water cannot get from one pore to another, it cannot flow through a substance.

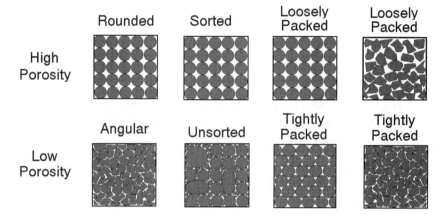

Factors affecting porosity.

PAINLESS TIP

A substance can have a lot of pore space and still be impermeable. Think of a block of Styrofoam. Its volume is mostly empty space, so it is very porous. But each pore space is like a little "bubble" separated from its neighbors by thin walls. Water cannot flow through Styrofoam because its pores are not interconnected.

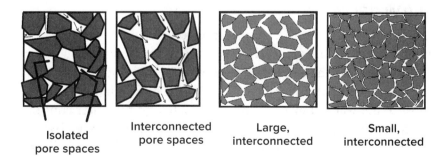

Factors affecting permeability.

BRAIN TICKLERS   Set # 5

1. The hydrosphere is about _____ percent fresh water.

2. _____ is water that seeps into the ground and fills the empty spaces between particles of rock or soil.

3. Three factors that affect the porosity of a sample of soil are the _____, _____, and _____ of the soil particles.

4. Which graph best represents the relationship between the particle size of a loose substance and the permeability of the loose substance?

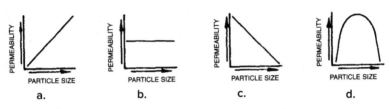

a.   b.   c.   d.

5. Which property is most nearly the same for the two soil samples shown below?

a. Porosity
b. Permeability
c. Infiltration rate

(Answers are on page 127.)

The water table

Gravity pulls water on Earth's surface down into the ground through interconnected pore spaces. Some of the water clings to the soil and plant roots near the surface forming a thin layer of moisture called the **soil moisture zone**. This is the layer from which plants get most of the water they need to survive. Most of the water continues to trickle downward through the ground until it is stopped by an impermeable substance. Unable to go any deeper, the water fills the pore spaces above the impermeable layer. The underground region in which all of the pore spaces are filled with water is the **zone of saturation**. The upper surface of the zone of saturation is called the **water table**. Below the level of the water table, the ground is saturated with water. Above the water table, the pore spaces are mostly filled with air. The underground region where the pore spaces are mostly filled with air is called the **zone of aeration**.

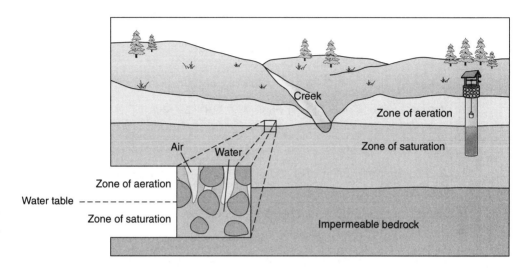

Zones of soil water and groundwater. The water table is the boundary between the zone in which pores are filled with air and the zone in which pores are filled with water.

Groundwater can leave the ground in a number of ways. It may evaporate and slowly diffuse out of the ground through the zone of aeration. The water table roughly follows the slope of Earth's surface. But if the surface of the ground drops below the level of the water table, the groundwater can seep out of the ground as a **spring**. The water from a spring can form streams, or collect in low spots forming ponds or lakes. If a stream wears down Earth's surface below the level of the water table, groundwater can seep directly into the stream and be carried away. Where the surface of the ground is at or just below the water table, the ground is always saturated with water and a swamp may form.

Groundwater can also be removed from the ground by digging a **well**. A well is a hole dug into the ground so that it penetrates the water table. Beneath the water table the well acts as a large pore space and fills with water. Then the water that fills the space can be pumped up to the surface through the well hole.

BRAIN TICKLERS   Set # 6

Write the letter of the term that best matches each definition.

a. Well

b. Groundwater

c. Spring

d. Water table

e. Zone of aeration

f. Zone of saturation

1. Water that seeps out of the ground where the surface dips below the water table

2. Water beneath Earth's surface found in pore spaces in rocks and soil

3. Underground region in which all of the pore spaces are filled with water

4. Underground region where the pore spaces are mostly filled with air

5. Surface where the zone of saturation and zone of aeration meet

6. A hole dug into the ground that penetrates the water table

(Answers are on page 127.)

Runoff

Runoff is precipitation or meltwater that does not evaporate or sink into the ground, but runs downhill along Earth's surface due to gravity. A number of characteristics of Earth's surface determine whether precipitation will become runoff. These include its permeability, slope, vegetation, and saturation with water.

The less permeable Earth's surface, the longer it takes water to infiltrate. When precipitation falls faster than it seeps into the ground, some of the water runs downhill before it has time to infiltrate. Therefore, as permeability decreases, runoff increases.

Water runs downhill faster on steep slopes than on gentle slopes. This allows less time for the water to infiltrate and more of the water runs off. Thus, as slope increases, runoff increases.

The stems of vegetation growing on Earth's surface block the path of water running downhill, causing it to run downhill more slowly.

This, in turn, allows the water more time to sink into the ground and thereby decreases runoff.

If heavy precipitation falls for a long time, pores can become saturated, or filled with water. What happens if you keep pouring water into a glass already filled with water? The glass overflows, of course. The same thing can happen if the pore spaces in the ground are already filled with water. When the ground is saturated, no more water can infiltrate and the excess water runs off.

Streams

A **stream** is a body of water that flows downhill due to gravity along a definite path. The place where a stream begins is called its **source**. Streams form as runoff from several slopes drains into low-lying areas between the slopes. These natural depressions fill up with water and form a puddle, a pond, or even a lake. As runoff continues, they fill and overflow. Then, the overflow continues flowing downhill along natural passageways or depressions in the surface to lower and lower levels. As the water flows along these passageways, it wears away the surface forming a clearly defined path, or **channel**. Once established, the same channel provides a pathway for all later runoff. The bottom of the channel is called the **streambed**. The sides of the channel are called the stream **banks**.

When two streams meet, they merge, and the smaller stream is known as a **tributary**. A stream grows larger as it collects water from more and more tributaries. The water in most streams ends up in a lake, a sea, or an ocean. The point at which the stream enters these bodies of water is called its **mouth**.

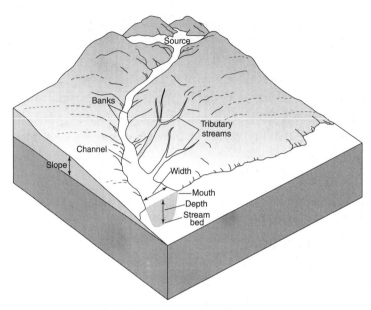

A typical stream with tributaries.

Stream flow

The water flowing in a stream, or **streamflow**, comes from two main sources: runoff and groundwater that seeps into the stream channel. Groundwater seeps into stream channels when the stream wears down Earth's surface so that the level of the streambed drops below the level of the water table. The groundwater that seeps into the stream is called its **base flow**. When you see water flowing in a stream during a dry spell when there is no runoff, it is most likely base flow.

When the moving water in a stream comes in contact with the air or the ground, friction causes the water to slow down. As a result, the water near the banks and streambed flows slower than the water in the deep, central part of the stream. Changes in the slope and shape of the channel also affect the speed of the stream.

The volume of water flowing past a given point in a stream in a given time is called the stream's **discharge**. In general, the greater the discharge of a stream, the faster the stream flows. A stream's discharge changes as the amount of precipitation changes. Discharge is greatest during wet times and least during dry times. During wet times, increased runoff can combine with base flow and cause the stream to overflow its banks, or **flood**.

PAINLESS TIP

Think of dumping a cup of water and a pail of water down a paved slope. The pail of water (the larger volume of water) would flow downhill faster.

Floods are often regular events. Thus, flooding is often predictable. Floods are most likely to occur during wet seasons or the spring melting of snow. Of course, flooding can also be sudden and unpredictable. Severe storms that bring heavy downpours or a sudden warm spell that melts snow quickly can all cause sudden flooding.

BRAIN TICKLERS   Set # 7

1. When rainfall occurs, the water will most likely become runoff if the surface of the soil is

   a. highly permeable.

   b. covered with vegetation.

   c. steeply sloped.

2. Which graph best shows the relationship between the slope of a streambed and the velocity of stream flow?

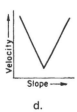

   a.   b.   c.   d.

3. The volume of water flowing past a given point in a stream in a given time is called the stream's _____.

4. Flooding always occurs when a stream's discharge is greater than the _____ of the stream's channel.

5. The length of the arrows in the two diagrams below represent the relative velocities of stream flow at various places in a stream.

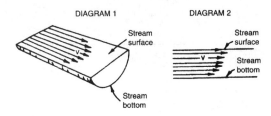

DIAGRAM 1   DIAGRAM 2

At which location is the velocity of the water flowing in a stream the greatest?

a. At the center along the bottom
b. At the center near the surface
c. At the sides along the bottom
d. At the sides near the surface

(Answers are on page 127.)

The Water Cycle

At Earth's surface, the lithosphere, hydrosphere, and atmosphere come in contact with one another and interact. Water moves among the hydrosphere, atmosphere, and lithosphere in a process called the water cycle.

Each day, water molecules at the surface of the oceans and other bodies of water receive energy from sunlight. The molecules heat up and trillions of tons of water evaporate and enter the atmosphere as water vapor. In the atmosphere, the water vapor is carried upward by rising warm air and is circulated by winds.

When the rising air cools to its dew point, the water vapor condenses back into liquid water forming tiny water droplets. These tiny water droplets suspended in air form clouds. When the water droplets become too large to remain suspended, gravity causes them to fall back to the surface as precipitation. Precipitation may fall on the oceans or the land. Water that falls on the oceans has completed its cycle and may evaporate back into the atmosphere, starting the process all over again.

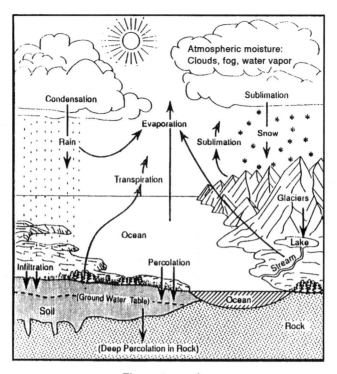

The water cycle.

A number of things may happen to precipitation that falls on land. It may become runoff and flow downhill into streams that eventually carry it back into the ocean or another body of water. It may infiltrate, become groundwater, and move slowly back to the oceans underground. Most of it seeps into rivers and streams that carry it back to the oceans.

Plants play a role in the water cycle as well. Water taken in by plant roots is transported to their leaves and released back into the atmosphere by a process called **transpiration**. In a forest, transpiration returns more than one-third of precipitation back to the atmosphere as water vapor. Some of the water may evaporate again to be carried farther over land, or may be blown back over the oceans.

Once the water is back in the oceans, the cycle begins all over again. There are many variations in the water cycle. Sometimes the water that evaporates over the oceans condenses and falls directly back into them as rain. Water falling on land may evaporate almost immediately, and in some cases precipitation evaporates before it ever reaches the ground. Other variations are possible, but all are part of the endless water cycle that causes global movements of water and changes in its form driven by energy from the Sun and the force of gravity.

In this way, the water cycle continually renews our supply of fresh water. Each day an estimated 15 trillion liters of water in the form of rain or snow falls on the United States alone. Earth is a closed system; therefore, the water in use today will eventually be recycled for use by a future generation. In theory, Earth's fresh water will never be exhausted. However, if long-lasting pollutants enter the basins and ground from which our water supply is drawn, water that is cleansed by evaporation is polluted as soon as it condenses and falls back into these polluted areas. Consider what happens if the ground is contaminated with nuclear waste, which remains radioactive for tens or even hundreds of thousands of years. As long as the ground remains radioactive, any water that infiltrates that ground will immediately become unusable. That is why, even though fresh water is a renewable resource, it is very important that we protect our sources of fresh water.

BRAIN TICKLERS   Set # 8

Base your answers to questions 1–4 on the diagram below, which shows the water cycle.

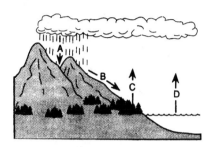

1. Water is returning to Earth's surface during step

   a. A.   **b.** B.   **c.** C.   **d.** D.

2. What is the main source of energy for the water cycle?

   a. The Moon   **b.** The Sun   **c.** Winds   **d.** Oceans

3. The primary source of most of the moisture for Earth's atmosphere is

   a. rivers and lakes.   **c.** oceans.

   b. melting glaciers.   **d.** groundwater.

4. Which arrow represents the process of transpiration?

   a. A   **b.** B   **c.** C   **d.** D

5. The flowchart below shows part of Earth's water cycle. Which process should be shown in place of the question marks to best complete the flowchart?

   Precipitation → runoff → ocean → ??? → water vapor

   a. Condensation   **c.** Evaporation

   b. Precipitation   **d.** Runoff

(Answers are on page 127.)

Wrapping up

- The hydrosphere is all of the water blanketing Earth's surface. It formed as Earth cooled and the water vapor in the atmosphere from comets and outgassing condensed and fell to the surface.

- Seawater is a complex mixture of dissolved inorganic matter, dissolved organic matter, and dissolved gases and particulates.

- The ocean floors include features such as the continental shelves, slopes, and rises; abyssal plains; mid-ocean ridges with rift valleys; and trenches.

- Ocean currents are caused by differences in the water's density and by global wind systems.

- Water seeps into Earth's surface through empty spaces between rock and soil particles by a process called infiltration.

- In order for water to infiltrate, Earth's surface must be both porous and permeable.

- Water that infiltrates and fills empty pores below Earth's surface is called groundwater.

- The water table is the boundary between the zone of aeration where pores are filled with air and the zone of saturation where pores are filled with water.

- Groundwater seeps out of Earth wherever the surface drops below the level of the water table forming springs, swamps, and feeding streams.

- Streams form when runoff overflows depressions and wears a channel in the land as it flows downhill.

- Earth's water cycles between the hydrosphere, atmosphere, and geosphere as energy from the Sun causes water to evaporate. Cooling in the atmosphere returns water vapor to the liquid or solid state. The water cycle constantly renews Earth's supply of fresh water.

Brain Ticklers—The Answers

Set # 1, page 104

1. hydrosphere
2. comets
3. atmosphere
4. oceans; ice caps and glaciers; freshwater lakes; rivers

Set # 2, page 107

1. d
2. b
3. a
4. e
5. c

Set # 3, page 108

1. dissolved inorganic matter; dissolved organic matter; dissolved gases
2. salinity
3. 35
4. a
5. a

Set # 4, page 112

1. cold; toward
2. bottom; warmer
3. cool; toward
4. c

Set # 5, page 116

1. three
2. Groundwater
3. shape; packing; sorting
4. a
5. a

Set # 6, page 118

1. c
2. b
3. f
4. e
5. d
6. a

Set # 7, page 121

1. c
2. a
3. discharge
4. volume
5. b

Set # 8, page 125

1. a
2. b
3. c
4. c
5. c

Chapter 5

The Lithosphere

The lithosphere, or "rock sphere," is the part of our planet made of rock, including the rock materials at Earth's surface and in all of its interior layers. Most rocks that form the lithosphere are made up of one or more substances called minerals. Earth's rocks and minerals are a rich source of natural resources ranging from fuels to building materials, metals, chemicals, and gems. Almost everything you have or use can be traced back to Earth's lithosphere. Understanding how rocks and minerals form helps us understand how Earth may have formed. It is also important to understand that Earth's natural resources are limited and should be used wisely.

Minerals

If you walk outside and pick up any Earth material—a rock, sand, soil, gravel, mud—you will hold minerals in your hand. Nearly all rocks are composed of one or more substances called minerals. But what exactly is a mineral and how is it different from a rock?

What is a mineral?

A **mineral** is a naturally occurring, inorganic, crystalline solid with a fixed chemical composition. If a substance does not have all of these characteristics, it is not considered a mineral. What do these characteristics mean?

Minerals are naturally occurring

Naturally occurring means the mineral formed as a result of natural processes in or on Earth. It was not manufactured by people or synthesized in a laboratory.

Minerals are inorganic

Inorganic substances are not alive, never were alive, and do not come from living things. Thus, amber (a tree resin in which insects are often found embedded), and the fossil fuels coal, petroleum, and natural gas are not true minerals. They were formed from organic substances, animals, or plants that once lived on Earth.

Minerals are crystalline

When the atoms or molecules that make up a mineral are joined in fixed positions as a solid, a definite pattern is formed. A solid having a definite, internal structural pattern is said to have a crystalline form. If the pattern is large enough to be seen with the unaided eye, the solid is called a **crystal**. The crystal form of minerals determines how they break apart and many other properties. For example, the mineral mica splits into thin, flat sheets because its molecules are arranged in thin, flat sheets.

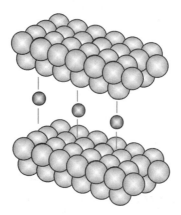

Diagram showing the crystalline pattern of mica.

Minerals have a fixed chemical composition

Minerals are either elements or compounds. **Elements** are substances that cannot be broken down into simpler substances by ordinary chemical means. Ninety or so different elements have been found occurring naturally on Earth, each with its own set of physical and chemical properties. Elements are made up of particles called atoms, and all atoms of an element are alike, but they are also different from the atoms of all other elements. Letters called symbols are used to represent the atoms of elements. For example, the symbol for the element oxygen is O and the symbol for the element silicon is Si.

Average Chemical Composition of the Hydrosphere, Atmosphere, and Lithosphere.

Element (symbol)	Crust		Hydrosphere	Troposphere
	Percent by Mass	Percent by Volume	Percent by Volume	Percent by Volume
Oxygen (O)	46.10	94.04	33.0	21.0
Silicon (Si)	28.20	0.88		
Aluminum (Al)	8.23	0.48		
Iron (Fe)	5.63	0.49		
Calcium (Ca)	4.15	1.18		
Sodium (Na)	2.36	1.11		
Magnesium (Mg)	2.33	0.33		
Potassium (K)	2.09	1.42		
Nitrogen (N)				78.0
Hydrogen (H)			66.0	
Other	0.91	0.07	1.0	1.0

The relative abundance of different elements in the crust of Earth's lithosphere. Notice that only eight elements make up more than 98% of the crust with the two most common elements being silicon and oxygen.

Some minerals can be found in the lithosphere as pure elements, such as copper and silver. But most elements are found chemically combined with other elements as **compounds**. In a compound, atoms of two or more elements join together in a fixed ratio in well-defined groups. The links that hold the atoms together are called **chemical bonds**. When linked by chemical bonds, each group of atoms forms a tiny particle, or **molecule**. For example, the mineral calcite is a compound of the elements calcium, carbon, and oxygen. It consists of molecules containing one atom of calcium, one atom of carbon, and three atoms of oxygen.

THE MINERALS IN YOUR SMARTPHONE

Of the 90 or so naturally occurring elements, smartphones can contain a whopping 70 different elements. In addition to copper, gold and silver for wiring, lithium and cobalt in the battery, and indium tin oxide in the touchscreen, the rare Earth metals also play a key role. Your phone couldn't vibrate without neodymium, terbium, and dysprosium. Terbium and dysprosium are also used to produce the colors of the display. Rare Earth metals are called rare not because they aren't plentiful, but because it's difficult to find them in concentrations high enough to be exploited. Every time we use one of these elements, that's it: we only have a finite amount of them, and there are no suitable replacements.

Earth's minerals are a treasure of almost unimaginable value. Unfortunately, minerals are considered a nonrenewable resource; that is, they may not form again within your lifetime or even a thousand lifetimes. Thus, Earth's mineral resources will not last forever. That is why it is so important that we conserve them by reducing waste and by reusing and recycling wherever possible.

PAINLESS TIP

Elements consist of **atoms** represented by **symbols**. **Compounds** consist of **molecules** represented by **chemical formulas**.

To show the makeup of the molecules of a mineral compound, a **chemical formula** can be written. Chemical formulas are made up of the symbols for the elements in the compound, each followed by a number telling how many atoms of that element are in a molecule of that compound. No number following a symbol means there is only one atom of that element in the molecule. The chemical formula for calcite is $CaCO_3$. The mineral quartz consists of molecules containing one atom of silicon and two atoms of oxygen. The chemical formula for quartz is SiO_2.

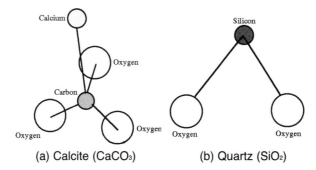

(a) Calcite (CaCO₃)   (b) Quartz (SiO₂)

Molecular models of calcite and quartz.

Every compound has its own distinct set of properties that are different from the properties of the elements of which it is composed. Therefore, every mineral, which is either an element or a compound, has both a definite composition and distinct properties all its own. This table shows the chemical names and formulas of some common minerals.

Some common minerals.

Mineral	Chemical Name	Chemical Formula
Calcite	Calcium carbonate	$CaCO_3$
Galena	Lead sulfide	PbS
Gypsum	Calcium sulfate-water	$CaSO_2 . 2H_2O$
Olivine (fosterite)	Magnesium silicate	Mg_2SiO_4
Potassium feldspar	Potassium aluminum silicate	$KAlSi_3O_8$
Pyrite	Iron sulfide	FeS_2
Quartz	Silicon dioxide	SiO_2

BRAIN TICKLERS   Set # 1

1. All of the atoms that make up an element are

    a. alike, but different from those of other elements.

    b. always spaced the same distance apart.

    c. different, but have the same mass as atoms of other elements.

    d. always moving at the same speed.

2. Elements chemically combine to form _____.

3. All minerals are

    a. solid.   c. elements or compounds.

    b. naturally occurring.   d. all of the above.

4. The most plentiful element in Earth's crust is

    a. quartz.   b. oxygen.   c. water.   d. silicon.

5. The mineral pyrite is made up of iron (Fe) and sulfur (S) atoms; it is considered to be a

    a. compound.   b. element.   c. mixture.

6. The chemical formula for the mineral hematite is Fe_2O_3. A molecule of hematite is made up of _____ atoms of the element iron (Fe) and _____ atoms of the element oxygen (O).

7. The pie graph below shows the elements comprising Earth's crust in percent by mass.

    Which element is represented by the letter X?

    a. Silicon   c. Nitrogen

    b. Lead   d. Hydrogen

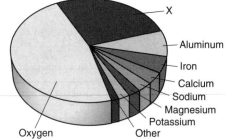

(Answers are on page 155.)

Mineral identification

Each mineral has a definite chemical composition—it is either an element or a compound. Therefore, every mineral has properties all its own. Because no two minerals have exactly the same set of properties, those properties can be used to identify minerals. Some properties that are commonly used to identify minerals because they are easy to observe are color, luster, streak, hardness, and cleavage.

Color is often the first property you notice about a mineral. However, color alone is not a reliable way to identify a mineral. Many minerals have almost the same color, whereas some minerals may occur in a variety of colors.

Luster is the way light reflects from a mineral's surface. Luster can be metallic or nonmetallic. Nonmetallic lusters can be described as glassy, brilliant, greasy or oily, waxy, silky, pearly, or earthy.

Streak is the color of the powder left when a mineral is rubbed against a hard, rough surface. A piece of unglazed porcelain, called a streak plate, is often used for this purpose. Although the color of a mineral may vary, the streak of a mineral is always the same. This makes streak a useful property for identifying a mineral. However, a mineral's streak is not always the same as the mineral's color. For example, pyrite has a brassy yellow color, whereas its streak is greenish black.

Hardness is a mineral's resistance to being scratched. The hardness of minerals is usually stated as a number using Mohs hardness scale.

Mohs Mineral Hardness Scale.

Talc	1
Gypsum	2
Calcite	3
Fluorite	4
Apatite	5
Feldspar	6
Quartz	7
Topaz	8
Corundum	9
Diamond	10

Approximate Hardness of Common Objects.

Fingernail	2.5
Copper penny	3.5
Iron nail	4.5
Glass	5.5
Steel file	6.5
Streak plate	7.0

Mohs hardness scale (1 is softest; 10 is hardest).

On this scale, ten common minerals are arranged in order from softest to hardest. On Mohs hardness scale, ten minerals are arranged in order from softest to hardest. See table on page 135. To find the hardness of a mineral sample, you determine which minerals the sample can scratch, and which minerals it cannot scratch. For example, halite will scratch gypsum or anything softer. It cannot scratch calcite or anything harder. The hardness of halite is between 2 and 3 (sometimes expressed as 2.5).

Cleavage is the tendency of a mineral to break along flat surfaces. These flat surfaces follow planes of weakness in their atomic structure. Cleavage surfaces often occur at very specific angles to one another and can be helpful in identifying a mineral. Some minerals do not break along smooth surfaces. They break unevenly, or **fracture**.

Because no single property can be used to identify all minerals, mineral identification is usually a process of elimination. As each property is observed, one discovers what a mineral is not, rather than what it is. Step-by-step, the possibilities are narrowed down until the identity of the mineral is determined.

BRAIN TICKLERS   Set # 2

1. Like all elements and compounds, a mineral has a definite
   _____ _____.

2. Each mineral has properties that are (like, different from) the properties of all other minerals.

3. The color of the powder left by a mineral after it has been scratched against an unglazed piece of porcelain is known as its
   _____.

4. The way a mineral shines is a property known as _____.

5. Which property of a mineral is tested by scratching it on a glass plate?

   a. Conductivity   b. Melting point   c. Density   d. Hardness

6. Which property would be most useful in identifying a sample of a mineral?

   a. Hardness   b. Size   c. Color   d. Smoothness

7. Certain minerals usually break along flat surfaces, whereas other minerals break unevenly. This characteristic is due to the

   a. luster of the mineral.

   b. age of the mineral.

   c. internal arrangement of the mineral's atoms.

   d. force with which the mineral is broken.

   Base your answers to questions 8 and 9 on the table showing the Mohs hardness scale on page 135.

8. Identify *one* mineral on the Mohs hardness scale that would be soft enough for an iron nail to scratch.

9. Explain why the property of hardness is usually better to use to identify a mineral instead of the color of the mineral.

   Base your answers to questions 10 through 12 on the table of mineral properties below.

| Mineral | Properties | | |
	Hardness	Streak	Reaction with Acid
calcite	soft	colorless or white	bubbles
chalcopyrite	hard	gray or black	rotten-egg smell
feldspar	hard	colorless or white	no reaction
galena	soft	gray or black	rotten-egg smell
graphite	soft	gray or black	no reaction
gypsum	soft	colorless or white	no reaction
hornblende	hard	gray or black	no reaction

10. Identify the mineral in the table that is hard, has a black streak, and has no reaction with acid.

11. Compared to the chalcopyrite, which property of galena is different?

12. Describe the test for determining the streak of most minerals.

(Answers are on page 155.)

Rocks

A **rock** is a naturally formed, nonliving Earth material that holds together in a firm, solid mass. Most rocks are made of lots of grains stuck together. The grains may be mineral crystals or tiny bits of broken rock, or even solid parts of once-living things. The grains may be small or large, tightly or loosely bound together, and may all be made of one mineral or may be a mixture of many different minerals.

Rocks are grouped into three "families" based upon how they were formed: igneous, metamorphic, or sedimentary. **Igneous** rocks form by the solidification of molten rock; **metamorphic** rocks form when heat, pressure, or chemical activity cause changes in existing rocks; and **sedimentary** rocks form by the chemical precipitation or the compaction and/or cementation of sediment. The composition and structure of a rock give us important clues as to how it formed.

Igneous rocks

Igneous rocks form when a molten mixture of minerals cools and turns into a solid. This molten mixture of minerals is called **magma** while it is beneath Earth's surface and **lava** when it escapes onto the surface. As the magma cools, molecules slow down, are attracted to one another, and settle into stable structures—molecules joined in fixed patterns, or mineral **crystals**.

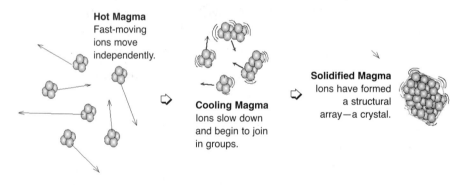

Hot Magma
Fast-moving ions move independently.

Cooling Magma
Ions slow down and begin to join in groups.

Solidified Magma
Ions have formed a structural array—a crystal.

Cooling and crystallization.

The rate at which molten rock cools off determines how large its mineral crystals can become. The longer the molten rock takes to cool, the more time the crystals have to form, and the larger they can grow. Therefore, the size of mineral crystals in an igneous rock are a clue to the speed at which it solidified, and that in turn is a clue to its origin.

PAINLESS TIP

To remember how the cooling rate affects crystal size, think:

Cooling **SLOW** lets crystals **GROW**; or **BIG** time, **BIG** crystals; **tiny** time, **tiny** crystals.

Intrusive and extrusive igneous rocks

Earth's surface is much colder than molten rock, so when lava is pushed out of a volcano, or *extruded*, it cools rapidly. Therefore, rocks that form from lava, or **extrusive rocks**, have tiny crystals that cannot be seen with the unaided eye. *Basalt*, a common volcanic rock, forms in this way. If cooling is *very* rapid, the rock may solidify before crystals have a chance to form. The result is a substance with no fixed structure—a glass. Volcanic glasses, such as *obsidian*, can be observed forming when lava flows into the ocean and cools very rapidly.

Crystals large enough to be seen by the unaided eye can form only if cooling is very slow. This occurs underground when magma pushes, or *intrudes*, into openings in underground rock and ends up blanketed by the surrounding rock. Rock is a poor conductor of heat. Magma surrounded by rock, like hot coffee in a thermos bottle, loses heat very slowly. Some underground pools of magma cool only a few degrees per century! As a result, rocks that form from magma, or **intrusive rocks**, have large, visible crystals like those seen in *granite*.

BRAIN TICKLERS   Set # 3

1. The main difference between rocks and minerals is that rocks are

   a. harder than minerals.

   b. usually mixtures so their composition is not always the same.

   c. found in the lithosphere, while minerals are not.

   d. crystalline, whereas minerals are not.

2. Rocks are classified as sedimentary, metamorphic, or igneous on the basis of the

   a. age of the rocks.

   b. way the rocks were formed.

   c. types of fossils the rocks contain.

   d. number of minerals found in the rocks.

3. Most igneous rocks form by which processes?

   a. Cooling and solidification

   b. Evaporation and condensation

   c. Erosion and deposition

   d. Compaction and cementation

4. Large crystal grains in an igneous rock indicate that the rock was formed

   a. near the surface.

   b. at a high temperature.

   c. from lava.

   d. over a long period of time.

(Answers are on page 155.)

Sedimentary rocks

Sedimentary rocks are all composed of sediments. **Sediments** are solid pieces of substances that have been carried along and then dropped by air, water, or ice. Most sediments are fragments of rock, but they could also be bits and pieces of plants and animals, or even molecules dissolved in water. Sediments are grouped and named by size as shown in the table below.

The Wentworth Scale.

Particle-Size Range (diameter, cm)	Particle Name	Sediment Composed of That Particle
Greater than 25.6	Boulder	Boulder gravel
6.4–25.6	Cobble	Cobble gravel
0.2–6.4	Pebble	Pebble gravel
0.006–0.2	Sand	Sand
0.0004–0.006	Silt	Silt
Less than 0.0004	Clay	Clay

How sedimentary rocks form

As layers of sediments pile up, they are changed into a solid rock by processes such as compaction and cementation. **Compaction** is the squeezing together of sediments due to the sheer weight of the layers on top of them, or by pressures exerted when the crust moves. Compaction forces the sediments closer together and they become more tightly packed. They also dry up as water is forced out of the spaces between sediment particles. As the water dries up, minerals dissolved in the water crystallize and help hold the rock together.

Cementation is the binding together of sediment particles by substances that crystallize or fill in spaces between loose particles of sediment. These substances literally act as cement, holding particles together in a solid mass.

Types of sedimentary rocks

Sedimentary rocks themselves can be grouped according to the type of sediment from which they form. **Clastic rock** is sedimentary rock that forms from broken bits and pieces of rocks and minerals eroded from the land and then deposited in layers that harden into rock. Clastic rocks are named according to the size of the rock fragments they contain. For example, sand-sized fragments form sandstone. Conglomerate is a clastic rock made up of fragments that are pebble sized or larger embedded in smaller sediments. **Chemical rock** is a sedimentary rock that forms from minerals that were once dissolved in water. Some form when water evaporates and leaves behind the dissolved minerals. Others form when chemical reactions cause minerals dissolved in the water to crystallize and settle through the water, or **precipitate**. The crystals settle to the bottom, forming layers that later harden into rock. Rock salt, rock gypsum, and dolostone are chemical sedimentary rocks. **Organic rocks** form from substances that were once part of, or made by, living things. For example, coal is an organic sedimentary rock that forms from compacted plant remains. Some limestones form when animals with shells die and the shells sink, building up in layers that harden into rock.

FOSSIL FUELS

Coal is formed from the remains of dead plants that were submerged in swamp environments. Over time, the plant remains were buried and subjected to the geological forces of heat and pressure over hundreds of millions of years. The result is coal—a black or brownish-black organic sedimentary rock with a high amount of carbon and hydrocarbons that is burned for fuel. Along with oil and natural gas, coal is one of the three most important fossil fuels. The main use of coal in the United States is the production of electricity.

Coal is called a fossil fuel because it forms from the fossilized remains of plants that lived millions of years ago. Fossil fuels are considered **nonrenewable** because they take millions of years to form. This means that fossil fuels are limited in supply and will eventually run out. That is why it is so important to develop renewable energy sources such as solar, wind, and water.

BRAIN TICKLERS Set # 4

1. Two processes that change sediments into sedimentary rock are _____ and _____.

2. Three types of sedimentary rocks are _____, _____, and _____.

3. Which process would form a sedimentary rock?

   a. Cooling of molten magma within Earth's crust

   b. Recrystallization of unmelted material within Earth's crust

   c. Cooling of a lava flow on Earth's surface

   d. Precipitation of minerals as the seawater evaporates

4. _____ sedimentary rock consists of broken fragments of rocks and minerals.

5. Which type of sedimentary rock is concrete most like? Explain your answer.

(Answers are on page 156.)

Metamorphic rocks

Metamorphic rocks get their name from the Greek words *meta*, meaning "change", and *morph*, meaning "form." Metamorphic rocks have literally been changed from their original form. Rocks change when they are exposed to an environment that is very different from the one in which they formed.

What causes metamorphic rocks to form

Three factors that cause changes in rock, or **metamorphism**, are heat, pressure, and chemical activity. Most metamorphism takes place deep underground where rock is exposed to high temperatures and pressures.

Heat

Earth's interior is very hot. The deeper you go below the surface, the higher the temperature rises. Rocks are heated when they get buried deep beneath Earth's surface or come in contact with hot substances such as magma. Metamorphism occurs at temperatures between 100°C and 800°C. Within these temperatures, a rock is still solid, but softened. Heating causes a mineral's molecules to move more rapidly and spread apart. This stretches and weakens the bonds that hold them together. Heating breaks some, but not all, of the bonds in the minerals that make up a rock. The minerals don't melt, but some of the atoms may rearrange themselves. These atoms join with others to form new minerals in the rock. This changes both the chemical composition and the structure of the rock.

Pressure

Pressure has the opposite effect; it forces the molecules in the mineral closer together. Pressure may be exerted on rock if it gets deeply buried and squeezed by the weight of the overlying layers. It may also be the result of movements of Earth's lithosphere that squeezes rocks together. When squeezed, the molecules in the rock rearrange into a more compact structure. The result is a denser, harder rock.

When grains recrystallize or are flattened under pressure, they form layers called **foliation**. Foliation can result from **mineral alignment** as pressure flattens or elongates mineral crystals in parallel layers. When minerals of different density recrystallize under pressure, they separate into layers like a mixture of oil and water. Light-colored minerals tend to be less dense than dark-colored ones. As the minerals separate into layers according to their density, they form a series of alternating light and dark layers called **banding**.

When sedimentary rocks are metamorphosed, layers in the rock may be distorted. Softened by the heat and squeezed by the pressure, the once flat layers become twisted and contorted. Such distortion is commonly seen in large-scale formations of metamorphic rock.

Some effects of pressure on rock particles are shown in the figure below.

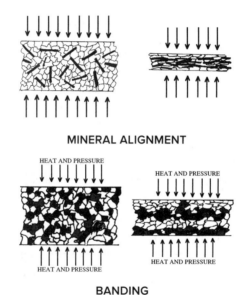

MINERAL ALIGNMENT

BANDING

Heat and pressure change a rock's structure.

Chemical activity

Rock can also be changed when hot, watery fluids rich in dissolved chemicals move through the rock. Fluids like this are given off by cooling magmas or may come from metamorphism taking place deep underground. As these fluids move through openings in a rock, their chemicals react with minerals in the rock to form new minerals.

Types of metamorphism

There are two main ways in which metamorphism takes place: **regional metamorphism** and **contact metamorphism**. Regional metamorphism occurs when rocks over a large region are exposed to high heat and pressure due to deep burial or movements within Earth's lithosphere. Contact metamorphism occurs when rocks are exposed to extreme heat as a result of coming in contact with magma or lava. Contact metamorphism is often seen in rocks found around the edges of igneous rock formations.

BRAIN TICKLERS   Set # 5

1. Metamorphism is best defined as the

   a. compaction and cementation of fragments of other rocks and minerals.

   b. precipitation of minerals dissolved in water.

   c. cooling and hardening of molten magma or lava.

   d. changing of rock by heat, pressure, and chemical activity.

2. Metamorphism may involve all of the following EXCEPT

   a. formation of new minerals.

   b. changes in grain size and shape.

   c. recrystallization.

   d. complete melting of the rock.

3. The type of metamorphism that results from the heat of magma or lava is

   a. foliation and banding.

   b. contact metamorphism.

   c. regional metamorphism.

   d. compaction and cementation.

4. Metamorphic rocks with mineral crystals arranged in parallel layers or alternating bands of light- and dark-colored minerals are

   a. intrusive.   b. volcanic.   c. clastic.   d. foliated.

5. Which characteristic of rocks tends to increase as the rocks are metamorphosed?

   a. Density

   b. Porosity

   c. Volume

   d. Number of fossils present

(Answers are on page 156.)

Rock Identification

Most rocks are mixtures; therefore, they cannot be identified in the same way as minerals. A single rock may consist of several minerals, each with its own color, luster, streak, hardness, density, and so on. For example, granite may contain white quartz (hardness 7), pink feldspar (hardness 6), and black mica (hardness 2.5–3). Granite varies in color and hardness from place to place within the rock. For this reason, the properties used to identify minerals are not as helpful when identifying a rock.

Two properties that are useful in identifying rocks are texture and mineral composition. **Texture** is the size, shape, and arrangement of the mineral crystals or grains in a rock. **Mineral composition** is simply the minerals that comprise the rock.

Identifying igneous rocks

Igneous rocks are composed of randomly scattered, tightly interlocking mineral crystals. They do not contain fossils because melting destroys fossils. The main differences between igneous rocks are the size of those crystals and the minerals of which they are composed. Most igneous rocks consist of a mixture of several of the following common minerals: quartz, feldspar, biotite, hornblende, pyroxene, and olivine.

The chart on page 147 summarizes the properties of the most common igneous rocks.

This chart can be used to make a very basic identification of an igneous rock. In the upper half of the chart, some of the most common igneous rocks are arranged by characteristics such as where they form, crystal size, and texture. In the middle are arrows indicating the ranges of color, density, and major elements in the rock. The lower half is a diagram that shows the percent mineral composition of these rocks.

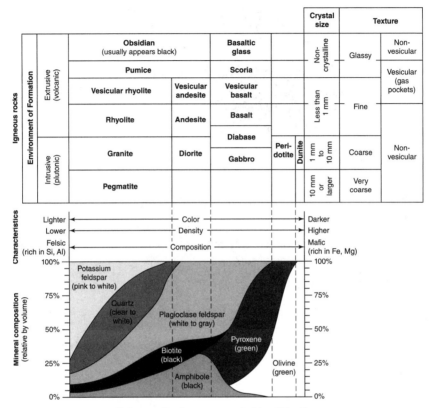

Scheme for igneous rock identification.

BRAIN TICKLERS   Set # 6

1. Which property would be most useful for identifying igneous rocks?

   a. Kind of cement holding grains together

   b. Mineral composition

   c. Mass of the rock sample

   d. Types of fossils present

2. Which igneous rock crystallized quickly near the surface of Earth, is light in color, and contains quartz and plagioclase feldspar?

   a. Granite   b. Gabbro   c. Basalt   d. Rhyolite

3. The diagram below represents a cross section of a coarse-grained igneous rock.

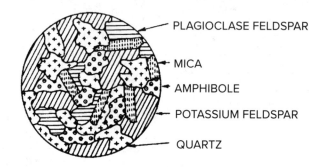

This rock is most likely to be

a. basalt.   **b.** gabbro.   **c.** rhyolite.   **d.** granite.

4. The diagram on the right represents the percentage by volume of each mineral found in a sample of basalt. Which mineral is represented by the letter X in the diagram?

a. Potassium feldspar   **c.** Quartz

b. Plagioclase feldspar   **d.** Mica

(Answers are on page 156.)

Identifying Sedimentary Rocks

Sedimentary rocks of all types have a number of unique character-istics. Sedimentary rocks form at or near Earth's surface *at normal temperatures and pressures*. Therefore, the minerals they contain are often quite different from those found in igneous or metamorphic rocks that form at high temperatures and/or pressures.

Sedimentary rocks *form in layers*. Most sediments on Earth's surface are moved by water. When sediments settle out of water, they form horizontal layers of particles. Each layer forms on top of the one that is already there. Therefore, each layer is younger than the one under it and older than the one on top of it, a principle called **superposi-tion**. By working out which rock layers that contain fossils are older and which are younger, the order in which living things evolved can be worked out.

Sedimentary rocks often contain distinctive features such as *ripple marks* created by waves in shallow water, *mud cracks* that formed when sediments dried out in the sun, and *footprints* of walking animals or the *tracks and trails* of crawling animals such as snails. Sedimentary rock may also contain **fossils**, the preserved remains of living things. Plants and animals often live in areas where sediments are being deposited. The remains of plants and animals that get buried in layers of sediment may then be preserved as fossils. These distinctive features are found only in sedimentary rocks. The melting that leads to the formation of igneous rocks and the high heat and pressure under which metamorphic rocks form destroys fossils and other traces of living things.

As shown in the chart below, texture and mineral composition are the main properties used to identify sedimentary rocks. The mineral composition of the sediment particles tells us where they came from, and the texture gives us clues about the processes that deposited the sediments.

Inorganic Land-Derived Sedimentary Rocks					
Texture	**Grain Size**	**Composition**	**Comments**	**Rock Name**	**Map Symbol**
Clastic (fragmental)	Pebbles, cobbles, and/or boulders embedded in sand, silt, and/or clay	Mostly quartz, feldspar, and clay minerals: may contain fragments of other rocks and minerals	Rounded fragments	Conglomerate	
			Angular fragments	Breccia	
	Sand (0.006 to 0.2 cm)		Fine to coarse	Sandstone	
	Silt (0.0004 to 0.006 cm)		Very fine grain	Siltstone	
	Clay (less than 0.0004 cm)		Compact; may split easily	Shale	
Chemically and/or Organically Formed Sedimentary Rocks					
Texture	**Grain Size**	**Composition**	**Comments**	**Rock Name**	**Map Symbol**
Crystalline	Fine to coarse crystals	Halite	Crystals from chemical precipitates and evaporites	Rock salt	
		Gypsum		Rock gypsum	
		Dolomite		Dolostone	
Crystalline or bioclastic	Microscopic to very coarse	Calcite	Precipitates of biologic origin or cemented shell fragments	Limestone	
Bioclastic		Carbon	Compacted plant remains	**Bituminous coal**	

Scheme for sedimentary rock identification.

BRAIN TICKLERS   Set # 7

1. A rock that contains fossil seashells is most likely

   a. igneous.

   b. sedimentary.

   c. metamorphic.

   d. volcanic.

2. Which sedimentary rock is most likely of organic origin?

   a. Limestone   b. Shale   c. Halite   d. Conglomerate

3. Which sedimentary rock could form as a result of evaporation?

   a. Conglomerate   b. Shale   c. Limestone   d. Sandstone

4. A sedimentary rock formed as the result of compaction and cementation of fragments of other rocks and minerals with a diameter between 0.006 and 0.2 centimeters is called

   a. breccia.   b. sandstone.   c. siltstone.   d. shale.

5. Bituminous coal is a sedimentary rock composed of

   a. cemented shell fragments.

   b. quartz, feldspar, and clay minerals.

   c. carbon from compacted plant remains.

   d. crystals of gypsum from chemical precipitates.

(Answers are on page 156.)

Identifying Metamorphic Rocks

Metamorphic rocks are also classified according to their texture and mineral composition. As you can see on the following page, meta-morphic rocks are divided into two main groups based on texture: foliated (layered) and nonfoliated. The foliation may be due to mineral crystals that lined up in parallel layers during recrystallization under pressure (mineral alignment) or because they separated by density into layers of dark and light minerals (banding). Then within each of these two groups, the rocks are classified by grain size. The column labeled "Composition" shows that foliated rocks share a common composition, whereas each nonfoliated rock has a distinctly different composition. The "Comments" column tells you the rock from which the metamorphic rock formed, or its parent rock.

Texture		Grain Size	Composition	Type of Metamorphism	Comments	Rock Name	Map Symbol
Foliated	Mineral alignment	Fine	Mica / Quartz / Feldspar / Amphibole / Garnet / Pyroxene	Regional (heat and pressure increases)	Low-grade metamorphism of shale	Slate	
		Fine to medium			Foliation surfaces shiny from microscopic mica crystals	Phyllite	
					Platy mica crystals visible from metamorphism of clay or feldspars	Schist	
	Banding	Medium to coarse			High-grade metamorphism; mineral types segregated into bands	Gneiss	
Nonfoliated		Fine	Carbon	Regional	Metamorphism of bituminous coal	Anthracite coal	
		Fine	Various minerals	Contact (heat)	Various rocks changed by heat from nearby magma/lava	Hornfels	
		Fine to coarse	Quartz		Metamorphism of quartz sandstone	Quartzite	
			Calcite and/or dolomite	Regional or contact	Metamorphism of limestone or dolostone	Marble	
		Coarse	Various minerals		Pebbles may be distorted or stretched	Meta-conglomerate	

Scheme for metamorphic rock identification.

BRAIN TICKLERS   Set # 8

1. Which rock is an example of a foliated metamorphic rock?

   a. Marble   c. Gneiss

   b. Quartzite   d. Granite

2. Slate forms when high pressure acts on and causes changes in the crystals of the sedimentary rock

   a. basalt.   c. granite.

   b. limestone.   d. shale.

3. Which metamorphic rock is composed of calcite or dolomite and formed by the metamorphism of limestone?

   a. Phyllite   b. Schist   c. Quartzite   d. Marble

4. Various rocks can be changed to the metamorphic rock hornfels by

   a. pressure due to regional metamorphism.

   b. contact metamorphism due to heat from nearby magma or lava.

   c. compaction and cementation of layers of sediments.

   d. precipitation of mineral crystals from evaporating seawater.

5. Which of the following lists the metamorphic rocks that formed as heat and pressure increased with depth?

   a. Gneiss–slate–schist   c. Slate–phyllite–schist

   b. Phyllite–gneiss–schist   d. Schist–phyllite–slate

(Answers are on page 156.)

The Rock Cycle

Where do rocks come from? You might say igneous rock comes from magma, sedimentary rock comes from sediments, and metamorphic rock comes from other rocks that have changed. But consider that magma is melted rock, sediments are broken rock, and there had to be an already existing rock to be metamorphosed. So, basically, rocks come from other rocks. Rocks are constantly changing from one type to another in a never-ending cycle called the *rock cycle*.

The rock cycle diagram

The diagram on page 153 shows the rock cycle. The outside circle shows the different forms in which rock matter can exist: magma, igneous rock, sediment, and so on. Arrows leading from one to another are labeled with the change that is taking place. Arrows inside the circle show some alternate paths in the rock cycle. For example, an igneous rock does not always break down into sediments. If buried, it may be metamorphosed. So, there is an arrow inside the circle leading from igneous rock to metamorphic rock.

The rock cycle diagram has no beginning and no end, but, of course, the rock cycle really did start somewhere. There is much evidence that Earth was originally totally molten. No solid rock existed, only magma. Thus, it is thought that the original rocks formed as this magma cooled, and the rock cycle began with igneous rocks.

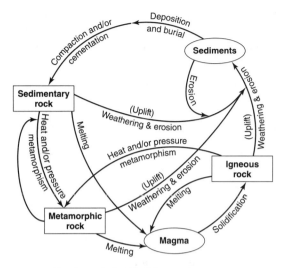

Rock cycle in Earth's crust.

BRAIN TICKLERS   Set # 9

Base your answers to the following questions on the *Rock cycle in Earth's crust* diagram.

1. Which types of rocks can be uplifted and eroded to form sediments?

   a. Igneous and metamorphic rocks only

   b. Metamorphic and sedimentary rocks only

   c. Sedimentary rocks only

   d. Igneous, sedimentary, or metamorphic rocks

2. Which two processes result in the formation of sedimentary rocks?

   a. Melting and solidification   c. Heat and pressure

   b. Compaction and cementation   d. Weathering and erosion

3. Metamorphic rocks can form from

   a. igneous rocks.   c. other metamorphic rocks.

   b. sedimentary rocks.   d. all of the above.

4. Which process leads to the formation of magma?

   a. Metamorphism   c. Cementation

   b. Melting   d. Solidification

5. What are two different series of events by which a sedimentary rock could form an igneous rock? Model your answers on the sample below, which shows one way in which an igneous rock could become a metamorphic rock.

   Sample: Igneous rock → heat and/or pressure → metamorphic rock

6. Basalt is a rock that is formed when magma from volcanic eruptions cools and solidifies. What class (type) of rock is basalt?

7. Identify *one* process that causes the metamorphism of an igneous rock into a metamorphic rock.

(Answers are on page 156.)

Wrapping up

- The lithosphere is Earth's rock layer.

- Rocks are composed of minerals.

- Minerals are naturally occurring, inorganic, crystalline solids with a fixed chemical composition.

- Minerals can be identified by their properties because no two minerals have exactly the same set of properties.

- Some properties that are commonly used to identify minerals because they are easy to observe are color, luster, streak, hardness, and cleavage.

- Rocks are grouped into three "families" based upon how they were formed: igneous, metamorphic, and sedimentary.

- Igneous rocks form by the solidification of molten rock.

- Molten rock is called magma while it is beneath Earth's surface and lava when it escapes onto the surface.

- Metamorphic rocks form when heat, pressure, or chemical activity causes changes in existing rocks.

- Sedimentary rocks form by the compaction and/or cementation of sediments or the crystallization of minerals dissolved in water.

- Two properties useful in identifying rocks are texture and mineral composition. Texture is the size, shape, and arrangement of a rock's mineral crystals or grains; mineral composition refers to the minerals that make up the rock.

- Rocks are constantly changing from one type to another in a never-ending cycle called the rock cycle.

Brain Ticklers—The Answers

Set # 1, page 134

1. a

2. compounds

3. d

4. b

5. a

6. 2; 3

7. a

Set # 2, page 136

1. chemical composition

2. different from

3. streak

4. luster

5. d

6. a

7. c

8. fluorite or calcite or gypsum or talc

9. Hardness is consistent from sample to sample, but the color can vary.

10. hornblende

11. hardness

12. Acceptable response include:

— drag the mineral across a piece of porcelain tile
— rub the mineral on an unglazed ceramic tile
— rub the mineral on a streak plate
— rub the mineral on a harder mineral
— crush the mineral into powder

Set # 3, page 139

1. b   2. b   3. a   4. d

Set # 4, page 142

1. compaction; cementation
2. clastic; chemical; organic
3. d
4. Clastic
5. Clastic; concrete is made up of cemented-together pieces of gravel and sand just as clastic rocks are made up of cemented-together rock fragments.

Set # 5, page 145

1. d   2. d   3. b   4. d   5. a

Set # 6, page 147

1. b   2. d   3. d   4. b

Set # 7, page 150

1. b   2. a   3. c   4. b   5. c

Set # 8, page 151

1. c   2. d   3. d   4. b   5. c

Set # 9, page 153

1. d   2. b   3. d   4. b

5. Series 1: sedimentary rock → melting → magma → solidification → igneous rock

   Series 2: sedimentary rock → heat and/or pressure → metamorphic rock → melting → magma → solidification → igneous rock

6. igneous *or* volcanic/extrusive

7. Acceptable responses include:

   — heat/heating
   — recrystallization of the minerals in the rock
   — compression/pressure
   — heat and/or pressure

Chapter 6

Weathering

Wherever rocks are exposed at Earth's surface, air, water, and living things cause them to change. They break down rocks into smaller pieces. Chemical reactions cause the minerals in rocks to change into new substances. Many of the processes that produce these changes result from exposure to the weather; therefore, the changes are called weathering. **Weathering** is the breaking down of rocks into smaller pieces by natural processes. Weathering processes are divided into two main types: physical and chemical.

Physical Weathering

Some of the changes caused by weathering are in form only. Weathering may break a large, solid mass of rock into loose fragments of all sizes and shapes but identical in composition to the original rock. Then only the size and shape of the rock have changed. Processes that break down rocks without changing their chemical composition are called **physical weathering**.

Some of the more important physical weathering processes include frost action, plant and animal action, abrasion, and changes in temperature.

PAINLESS TIP

Physical weathering = physical changes only (e.g., changes in size or *shape*).

Composition before = composition after

Frost action

One of the most important physical weathering processes is caused by water freezing in cracks or pores in rocks, or **frost action**. Frost action is the result of an unusual property of water. Most liquids expand when heated and contract when cooled. At first this is also true of water. Water behaves normally and contracts as it cools until it reaches 4°C. Then, amazingly, water stops contracting and begins to expand slowly as it is cooled. When water finally freezes into ice at 0°C it expands dramatically—increasing in volume by 9 percent. And when freezing water expands it can exert huge forces on anything confining it. Freezing water can exert forces measuring tens of thousands of pounds per square inch!

How does this unusual property of water break down rocks? Liquid water, in the form of rain, melting snow, or condensation, seeps into any cracks or pores in rock. If temperatures then drop below 0°C, the liquid water in the cracks freezes into ice. The expanding ice exerts tremendous pressure against the confining rock. Acting like a wedge, it widens and deepens the crack. When the ice thaws, the water seeps deeper into the crack. When it refreezes, the process is repeated. In this way, the alternate freezing and thawing of water, or *frost action*, breaks apart rocks.

Frost Wedging

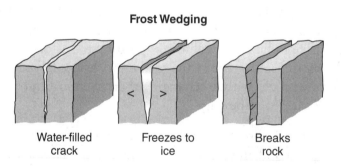

| Water-filled crack | Freezes to ice | Breaks rock |

Frost action is particularly effective where bedrock is directly exposed to the atmosphere, water is present, and the temperature fluctuates frequently above and below the freezing point of water. This is often the case during the winter in temperate climates. It can also occur on mountain tops and at high elevations in spring or fall. Daytime temperatures rise above freezing causing snow and ice to melt, only to drop below freezing again at night, producing frost action.

As you might expect, frost action is almost nonexistent in places like Florida and Hawaii. However, through much of the northeastern United States it is probably the main weathering process. Throughout the world, it is probably the most important physical weathering process.

PAINLESS TIP

Frost action is sometimes called frost wedging because the ice acts as a wedge to split rock apart.

Plant and animal action

Plants and animals also weather rock. When a rock develops cracks, small particles of rock and soil get washed into them by rain or blown in by wind. If a seed finds its way into the crack, it can germinate and begin to grow. As the plant grows, its roots thicken and press against the sides of the crack. Similar to the expansion of ice in frost action, the growing roots widen and extend the crack. Eventually the rock is broken apart. The roots of tiny plants like mosses and lichens produce a rock-dissolving acid as they grow and decay. This further speeds the breakdown of the rocks.

Animals also help weather rock. Burrowing animals, such as ants, termites, earthworms, woodchucks, and moles, speed up weathering by bringing fresh rock particles to the surface. Their burrows allow

Physical weathering by plant action.

air and water to penetrate the surface and weather the underlying bedrock. Charles Darwin once calculated that on 1 acre of land earthworms bring as much as 10 metric tons of rock grains to the surface each year. Exposed to the atmosphere again, these grains of rock are further broken down by weathering.

Humans also help weather rock. Rock quarrying, strip mining, and blasting to build roads are just a few of the human activities that break up rock and expose vast amounts of fresh rock to weathering processes.

Abrasion

Rocks can also be broken down by **abrasion**, or rubbing against each other. Rock abrasion occurs mainly when fragments are being carried along by agents of erosion. For example, as rock fragments are carried along by winds or water, they collide and rub against each other. This breaks off smaller pieces from the corners and sharp points that stick out from the fragment. As a result, the fragments become smaller and more rounded. The fragments also abrade the bedrock as they bounce and slide along its surface.

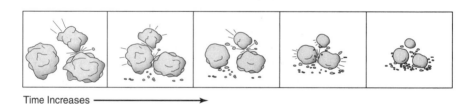

Time Increases ⟶

Physical weathering by abrasion.

Changes in temperature

Rocks are often exposed to large temperature changes. As rocks heat up during the day they expand; as they cool off at night they contract. You might expect this constant change to cause rocks to crack and break up. However, experiments have indicated that this is not generally the case. Only extreme temperature changes such as those resulting from forest and brush fires cause rocks to crack or flake off at the surface.

BRAIN TICKLERS   Set # 1

Write the letter of the term that best matches each definition.

1. Physical weathering

    a. The breaking down of rocks by collisions or rubbing together.

2. Abrasion

    b. Rock breaks along cracks into which roots of plants have grown.

3. Frost action

    c. The breaking down of rocks into smaller pieces without changing their chemical composition

4. Plant action

    d. The repeated freezing and thawing of water

5. Which is the best example of physical weathering?

    a. The rusting of an iron nail

    b. The cracking of rock caused by the freezing and thawing of water

    c. The chemical reaction of limestone with acid rain

    d. The formation of a sandbar along the side of a stream

6. Which property of water makes frost action a common and effective form of weathering?

    a. Water dissolves many Earth materials.

    b. Water expands when it freezes.

    c. Water cools the surroundings when it evaporates.

    d. Water has a density of 1 gram per cubic centimeter.

7. Which weathering process brings buried rock particles back to the surface where they can be further weathered?

    a. Pressure unloading   c. Frost action

    b. Plant action   d. Animal action

8. Four quartz samples of equal size and shape were placed in a stream. Which of the four quartz samples below has most likely been transported farthest in the stream?

a.   b.   c.   d.

(Answers are on page 177.)

Chemical Weathering

Chemical weathering breaks down rocks by changing their chemical composition. When rocks are exposed at Earth's surface, they come into contact with substances that were not present where they formed. The minerals in rock may react with these substances to form new substances with properties different from those of the original minerals. The new substances are usually softer, and this weakens the structure of the rock. As a result, the rock either falls apart or is more easily broken down by physical weathering. The substances that cause most of the chemical weathering of rocks are oxygen, water, and carbon dioxide.

Oxygen

Oxygen causes chemical weathering by combining with other substances in a chemical reaction called **oxidation**. Oxygen reacts most readily with minerals containing iron such as the minerals magnetite, pyrite, hornblende, and biotite. When oxygen combines with iron, iron oxides like rust are formed. The minerals hematite (Fe_2O_3) and magnetite (Fe_3O_4) are common examples of different kinds of iron oxides.

When oxygen combines with iron, the chemical bonds between the iron and other elements in the mineral compound of a rock are broken. This weakens the rock's structure and it can then be more easily broken down by other weathering processes. Other substances in rock, such as aluminum or silicon, can also combine with oxygen. The effects are much the same: forming an oxide weakens the structure of the rock and the rock is more easily broken down.

PAINLESS TIP

Oxidation—chemical reaction in which oxygen combines with another substance.

When thinking about how oxidation weakens a rock, compare the strength of a shiny, new iron nail with one that has oxidized into rust.

Water

Most of Earth's surface is covered with water. Trillions of gallons of water fall to the ground as rain every day. With so much water around, it is not surprising that water is an important agent of chemical weathering.

Water can chemically weather rock in several ways. First, water can weather rock simply by dissolving minerals—a process called **solution**. Many minerals dissolve in water to some extent. Water will slowly dissolve these minerals out of a rock, leaving behind empty spaces. This exposes the surrounding mineral grains to further weathering. Sometimes the rock's structure is so weakened by the empty spaces created that it just crumbles. Once dissolved in water, the dissolved minerals may react with each other to form new substances. If these new substances are not soluble in water they precipitate out.

PAINLESS TIP

Water is called the "universal solvent" because so many substances dissolve in water.

Second, water can chemically react with some minerals to form new minerals. For example, **hydration** is a chemical reaction in which water combines with another substance. When water combines with the mineral anhydrite, a new mineral is formed—gypsum.

Hydration of anhydrite to form gypsum:

$$CaSO_4 \ + \ 2H_2O \ \rightarrow \ CaSO_4 \cdot 2H_2O$$

Anhydrite   water   gypsum

Water can also break apart mineral molecules and then combine with the parts in a reaction called **hydrolysis**. Hydrolysis causes most minerals to swell and crumble. One example is the hydrolysis of feldspar to form kaolinite.

Hydrolysis of feldspar to form kaolinite:

$$4KAlSi_3O_8 + 4H^+ + 2H_2O \rightarrow Al_4Si_4O_{10}(OH)_8 + 8SiO_2$$

| Potassium feldspar | hydrogen ions | water | kaolinite | silica |

Carbon dioxide

Carbon dioxide as a gas has almost no effect on rocks. But when carbon dioxide comes in contact with water, a chemical reaction takes place. Water combines with carbon dioxide to form **carbonic acid**.

Formation of carbonic acid:

$$CO_2 + H_2O \rightarrow H_2CO_3$$

Carbon dioxide + water → carbonic acid

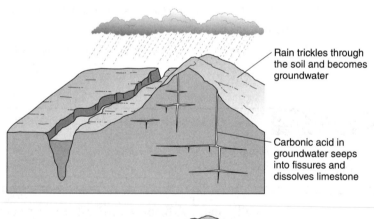

Rain trickles through the soil and becomes groundwater

Carbonic acid in groundwater seeps into fissures and dissolves limestone

Carbonic acid slowly enlarges fissures in limestone, forming caverns

Formation of caverns.

Carbonic acid is not a very strong acid, but when it comes in contact with a rock, it can dissolve some minerals, weakening the rock's structure. Carbonic acid can completely dissolve the mineral calcite found in limestone. When rainwater and groundwater containing carbonic acid seeps into limestone bedrock it dissolves the calcite and forms large underground spaces called **caverns**.

Other chemical factors

In addition to carbonic acid, there are other naturally occurring acids that attack rocks and minerals. Some are produced by lightning or decay of organic material. Others are produced as waste products of certain plants and animals. These acids dissolve in rainwater and seep down through the soil to bedrock; they dissolve some rocks and cause others to crumble.

Human activities have also added acids to the environment. Many of the waste gases given off by factories, homes, and automobiles contain substances that react with water to form very strong, reactive acids. Near some large cities and factories rainfall contains so much acid that it is called **acid rain**. Acid rain speeds up the weathering of rock and damages plant and animal life. It also corrodes structures made by humans.

BRAIN TICKLERS   Set # 2

1. _____ weathering changes the composition of minerals and _____ weathering does not.

2. Iron combines with _____ to form rust in a reaction called _____.

3. Carbon dioxide can chemically combine with water to form _____ acid.

4. Which of the following is an example of chemical weathering?

   a. Freezing of water in the cracks of a sandstone walkway

   b. Abrasion of a streambed by tumbling rocks

   c. Grinding of the mineral talc into powder

   d. Dissolving of limestone bedrock by acid rain

5. Chemical weathering will occur most rapidly if rocks are exposed to the

   a. hydrosphere and lithosphere.

   b. mesosphere and thermosphere.

   c. hydrosphere and atmosphere.

   d. lithosphere and atmosphere.

6. Underground caverns are formed by the action of carbonic acid on

   a. limestone bedrock.

   b. grains of feldspar.

   c. quartz sand.

   d. iron-rich minerals.

   Base your answers to questions 7 and 8 on the information below and your knowledge of science.

   A student performed an experiment in which 10 mL of a strong acid was placed on a sample of limestone. Bubbles formed where the acid touched the limestone. After 20 minutes, the bubbling stopped and the surface of the limestone appeared unchanged.

7. Identify *one* observation that shows a chemical reaction occurred between the acid and the limestone.

8. Explain why limestone buildings are weathered by acid rain even though the limestone sample in this experiment did *not* appear to be changed by the strong acid.

(Answers are on page 177.)

Factors that Affect Weathering

There are many factors that affect the rate at which a rock weathers. Among the more important factors are climate, mineral composition, particle size, exposure, and time.

Climate

Climate is the single most important factor affecting weathering. Warm, moist climates favor chemical weathering because reactions tend to occur at a faster rate as temperature increases; many reactions require water. Warm, moist climates also favor the growth of plants and animals that contribute to the weathering of rock. Cold climates

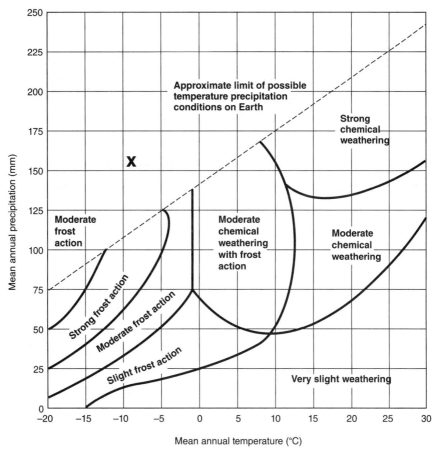

The effects of climate on weathering.

favor physical weathering, principally frost action. Here again, water is a key factor. Without water you cannot form ice and frost action cannot occur. Frost action is most intense in moist climates with wide temperature changes that lead to alternate freezing and thawing. For these same reasons, dry climates experience very little weathering. The chart above shows the types of weathering that occur in different types of climates.

Mineral composition

The weathering rate also depends on a rock's mineral composition. A rock's mineral composition determines its physical and chemical properties. Rocks composed of minerals that react readily with acids, water, or oxygen will weather more rapidly than those composed of minerals that are less reactive.

Softer rocks will abrade more readily than harder rocks. Solid, crystalline rocks have fewer openings into which water can penetrate than rocks composed of cemented-together particles. Therefore, mineral composition determines a rock's resistance to weathering.

Mineral Name	Relative Resistance to Chemical Weathering
Halite	Very low
Gypsum	Very low
Pyrite	Low
Calcite	Low
Dolomite	Low
Olivine	Moderately low
Pyroxene	Moderate
Plagioclase	Moderate
Hornblende (Amphibole)	Moderate
Biotite	Moderate
Potassium feldspar	Moderately high
Muscovite	High
Quartz	Very high

Particle size

Particle size also affects the rate of weathering because weathering can only affect exposed surfaces. Since a given volume of small particles has more surface area than the same volume of large particles, small particles tend to weather faster than large particles.

Size is also a factor in what happens to rock fragments after they are produced by weathering. Small particles may be transported to a new location that has a different climate. They may be carried to a body of water and may be exposed to water containing other mineral compounds.

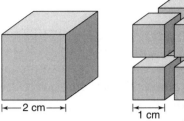

This cube has a surface area of **24 cm²**

This is the same cube divided into eight smaller cubes. Each of the eight small cubes has a surface area of 6 cm² for a total surface area of **48 cm²**.

As particle size decreases, total surface area increases.

The effects of particle size on total surface area and weathering.

Exposure

The more exposed rocks are to air, water, and living things, the faster they weather. Soil, ice, and vegetation can cover a rock and thereby decrease its exposure to weathering. The weathered surface of a rock itself can shield the rock beneath it, slowing the rate of weathering. The slope of the land also affects exposure. On steep slopes, loose materials move downhill due to gravity or are carried downhill by erosion. This continually exposes fresh rock.

Exposure also affects each individual piece of rock produced by weathering. Flat sides have one surface exposed to weathering. Corners and edges can have two or more surfaces exposed to weathering. Therefore, the corners of a piece of rock weather away fastest and flat surfaces weather away slowest. Over time this causes the piece of rock to become rounded.

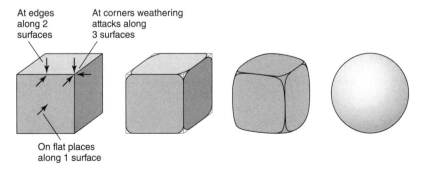

At edges along 2 surfaces

At corners weathering attacks along 3 surfaces

On flat places along 1 surface

Weathering causes particles to become rounded.

Time

Time is the most basic factor affecting weathering. The longer a rock is exposed to weathering processes, the more it is broken down. Weathering is a slow process. Even easily weathered rocks may take hundreds of years to be completely broken down. More resistant rocks may take millions of years to weather away. Eventually, though, all rocks exposed at Earth's surface are completely broken down by weathering processes.

BRAIN TICKLERS   Set # 3

1. The weathering of rocks and minerals is most affected by

   a. topography.

   b. longitude.

   c. altitude.

   d. climate.

2. Four samples of rock with identical composition and mass were cut as shown in the diagrams below. When subjected to the same chemical weathering, which sample will weather at the fastest rate?

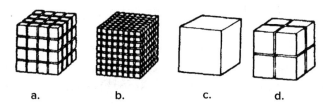

   a.   b.   c.   d.

Answer questions 3–6 based on the graph below, which was prepared from the results of a study of four different types of cemetery stones. The graph shows the relationship between the ages of four cemetery stones and the percentage of each stone that had weathered away.

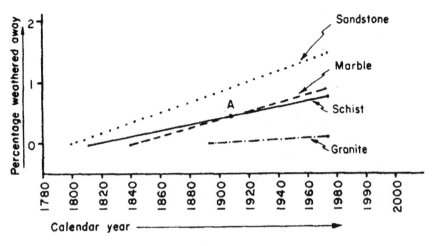

3. Which rock was found to have been exposed to weathering for the greatest number of years?

   a. Granite   c. Marble

   b. Schist   d. Sandstone

4. In this study, which rock was the most resistant to weathering?

   a. Marble   c. Granite

   b. Schist   d. Sandstone

5. What total percentage of the schist should have weathered away by the year 2000?

   a. 1.0 percent   c. 0.5 percent

   b. 2.0 percent   d. 1.5 percent

6. Point A on the diagram represents the time at which

   a. equal percentages of the marble and schist had weathered away.

   b. the marble and schist weathered away at the same rate.

   c. climatic conditions changed the weathering rates.

   d. industrial pollutants changed the weathering rates.

To answer questions 7–9, refer to the diagram on page 167 that shows the main type of weathering that occurs in various climate conditions.

7. Which climatic conditions would produce very slight weathering?

   a. A mean annual temperature of 25°C and a mean annual precipitation of 100 mm

   b. A mean annual temperature of 15°C and a mean annual precipitation of 25 mm

   c. A mean annual temperature of 5°C and a mean annual precipitation of 50 mm

   d. A mean annual temperature of −5°C and a mean annual precipitation of 50 mm

8. Why is no frost action shown for locations with a mean annual temperature greater than 13°C?

   a. Almost no freezing takes place at these locations.

   b. Large amounts of evaporation take place at these locations.

   c. Almost no precipitation falls at these locations.

   d. Large amounts of precipitation fall at these locations.

9. There is no particular type of weathering or frost action given for the temperature and precipitation values at the location represented by the letter X. Why is this the case?

   a. Only chemical weathering would occur under these conditions.

   b. Only frost action would occur under these conditions

   c. These conditions create both strong frost action and strong chemical weathering.

   d. These conditions probably do not occur on Earth.

10. In which type of climate does chemical weathering usually occur most rapidly?

   a. Hot and wet   c. Hot and dry

   b. Cold and wet   d. Cold and dry

(Answers are on page 177.)

Products of Weathering

Weathering processes have been attacking rocks at Earth's surface since they first formed. Whether these weathering processes are physical or chemical, the result is the same—solid rock is broken into fragments.

Sediments

The fragments or particles of rock produced by weathering are called **sediments**. Sediments are named by their size. See the table below.

Sediment sizes.

Particle Diameter	Name
> 25.6 cm	BOULDERS
6.4 cm – 25.6 cm	COBBLES
0.2 cm – 6.4 cm	PEBBLES
0.006 cm – 0.2 cm	SAND
0.0004 cm – 0.006 cm	SILT
0.00001 cm – 0.0004 cm	CLAY

Soil

Over time, weathering and plant growth change exposed rock and sediments into **soil**, a mixture of weathered rock, water, air, bacteria, and decayed plant and animal material (humus). Soil may form directly from the bedrock beneath it or from sediment that was carried to the location by erosion. As a soil forms, the processes of weathering and plant growth cause recognizable layers, or **horizons**, to form in the soil.

A soil that has been forming long enough to have developed distinct horizons is called a mature soil. It may take hundreds to thousands of years to develop a mature soil. The diagram on page 174 shows a cross section of the development of a mature soil from surface to bedrock and a description of each horizon.

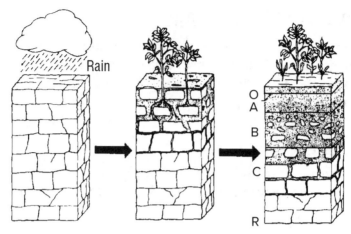

Development of a mature soil.

Horizon O – Organic. Leaf litter, twigs, roots, and other organic material lying on the surface of the soil.

Horizon A – Topsoil. Loose and crumbly with varying amounts of organic matter; generally the most productive layer of soil.

Horizon B – Subsoil. Usually light colored, dense, and low in organic matter.

Horizon C – Parent material. The partly weathered bedrock or transported sediments from which soil forms.

Parent material (R) – Bedrock. The solid rock that underlies the soil.

SOIL: IT ISN'T JUST DIRT!

You may think it's just dirt, but soil is a natural resource as precious as gold. In a way, soil is essential for life. Soils provide the medium for plant growth—the basis of most food chains. People depend on soil to grow food, for feed for animals, for fibers, and for fuel.

But soils do a lot more than that. Soils are a habitat for a whole variety of living things, from bacteria and fungi to insects, worms, and burrowing animals. Soils act as a filtration system for surface water and store water, which helps the environment bounce back after floods and drought. Soils also store carbon and help maintain atmospheric gases.

Many consider soil a nonrenewable resource because even under the best conditions, it takes natural processes about 500 years to develop an inch of fertile topsoil. To grow good crops takes about 6 inches of topsoil. That means it takes 3,000 years to develop enough topsoil to grow good crops. When soils that took thousands of years to develop are lost or degraded, they can't be recovered in a human lifetime. Since 1950, one-third of the world's arable land (land that can be used to grow crops) has been lost. Every year the United States loses billions of tons of soil. That is why the conservation and preservation of our existing soil is essential for food security and to ensure a sustainable future.

BRAIN TICKLERS   Set # 4

1. The formation of rock fragments and soil is most likely the result of

   a. weathering.   c. convection cells.

   b. gravity.   d. hazardous weather.

2. _____ is a mixture of weathered rock, water, air, bacteria, and decayed plant and animal material.

3. The horizontal layers or zones in a mature soil are known as

   _____.

4. The mineral and rock fragments that result from the breaking down of rock by weathering processes are known as _____.

5. Frost action breaks loose a fragment of rock from a cliff and it falls to the ground. The rock fragment is measured and found to be 15 centimeters in diameter. This fragment would best be classified as a

   a. boulder.   c. cobble.

   b. pebble.   d. sand grain.

6. An experiment is described below.

   A large field at the base of a mountain becomes flooded when heavy rains in the mountains cause a stream to overflow. Each time the flooding occurs, more soil washes away. The owners of the land want to perform an experiment to see if different types of plants can help reduce the soil erosion. They choose five areas of ground that are the same size and the same distance from the stream, have the same slope and the same kind of soil, and receive the same amount of sunlight. The type of plant planted in each area is different for each of the five areas. Measurements of soil erosion will be made each time flooding occurs. The results will be compared after six months.

Which hypothesis is being tested in this experiment?

a. Soil erosion is affected by the strength of the wind.

b. Flooded areas have greater soil erosion than areas that are not flooded.

c. Some types of plants reduce soil erosion more than others.

d. Some types of soil are more easily eroded.

(Answers are on page 177.)

Wrapping up

- Weathering is the breaking down of rocks into smaller pieces by natural processes.

- Physical weathering processes break down rocks without changing their chemical composition.

- The most important physical weathering process is frost action. Other physical weathering processes include plant and animal action, abrasion, pressure unloading, and exfoliation.

- Chemical weathering breaks down rocks by changing their chemical composition.

- Most of the chemical weathering of rocks is caused by oxygen, water, and carbon dioxide.

- Climate is the most important factor affecting weathering; mineral composition, particle size, exposure, and time also affect weathering.

- The fragments or particles of rock produced by weathering are called sediments.

- The end product of weathering is soil, a mixture of weathered rock, water, air, bacteria, and decayed plant and animal material (humus).

Brain Ticklers—The Answers

Set # 1, page 161

1. c	3. d	5. b	7. d
2. a	4. b	6. b	8. d

Set # 2, page 165

1. Chemical; physical

2. oxygen; oxidation

3. carbonic

4. d

5. c

6. a

7. Bubbles formed. *or* A gas formed.

8. Acceptable responses include, but are not limited to:

  — Acid rain damage takes years, not minutes.

  — Acid rain takes more than 20 minutes to damage buildings.

  — The change in the limestone in the experiment was too small to see.

Set # 3, page 170

1. d	3. d	5. a	7. b	9. d
2. b	4. c	6. a	8. a	10. a

Set # 4, page 175

1. a	3. horizons	5. c
2. Soil	4. sediments	6. c

Chapter 7

Erosion and Deposition

After weathering has broken down rock into smaller pieces, or sediment, other processes then pick up and move them. Any natural process that picks up sediment and moves it from place to place is called **erosion**. Together, weathering and erosion wear down Earth's surface. Weathering breaks down solid rock and erosion carries away the pieces. This exposes fresh rock to weathering and the cycle repeats itself.

Eventually, though, the sediment that was picked up and moved by erosion is dropped in a new place—a process called **deposition**. Deposition also changes the shape of Earth's surface. When sediments are deposited, they build up Earth's surface and create new landforms. Weathering, erosion, and deposition are all part of an unending interplay between the forces that wear down and forces that build up Earth's surface.

PAINLESS TIP

Deposition Drops off sediment; erosion picks up and carries away sediment.

Agents of Erosion

In order to move anything, a force is needed. The main driving force behind erosion is gravity. Gravity can move sediments by acting on them directly. For example, when a piece of rock is broken loose from a cliff by weathering, it falls because gravity pulls on it directly. Gravity can also move sediments by acting on them indirectly, through agents of erosion. For example, gravity pulls on the water, causing it to move downhill in streams. The moving water, in turn, can exert a force on sediments causing them to move. A substance set in motion by a driving force that can transport sediment is called an **agent of erosion**. Agents of erosion include moving water (streams, waves, and currents), moving ice (glaciers), and moving air (wind).

You may wonder how gravity causes air to move. The pull of gravity causes denser air to sink and less dense air to float upward. This creates the convection currents that cause winds, and winds can exert a force on sediments causing them to move. Winds can also exert forces on water surfaces creating waves and currents that can also pick up and carry sediments.

BRAIN TICKLERS   Set # 1

1. _____ is any process that picks up sediment and moves it from one place to another.

2. _____ is any process by which transported sediment is dropped in a new place.

3. _____ is the main driving force behind all erosion.

4. A substance set in motion by a driving force that can pick up and transport sediment is called an _____.

5. Name three common agents of erosion: _____, _____, and _____.

(Answers are on page 206.)

PAINLESS TIP

When thinking about erosion, think **GGWW**—**G**ravity, **G**laciers, **W**ater, **W**ind.

Moving water, ice, and air move in different ways and at different speeds and carry different sizes and amounts of sediment. Therefore, each of these agents of erosion produces distinctive changes in the sediment it transports and creates characteristic surface features and landscapes.

Gravity

Imagine a book resting on a desk. While the desk is flat, the book does not move. Gravity, pulling down on the book, holds it in place. What happens, though, if one end of the desk is slowly tilted upward? After a while, the book slides downhill. This is because on a slope, gravity acts as if it has two parts. One part pulls down at right angles to the surface and the other pulls downhill parallel to the surface. As the slope gets steeper, more and more gravity acts in a downhill direction. When the downhill force is greater than the force holding the book against the surface, the book will move downhill.

Mass movements are downhill movements of rock and soil under the direct influence of gravity. On most slopes, some kind of mass movement is going on all the time. The rate at which it occurs depends on the steepness of the slope. This is because the steeper the slope, the greater the force of gravity acting in a downhill direction against friction.

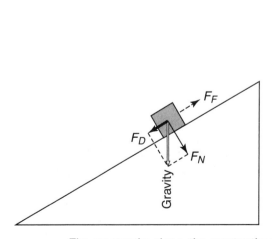 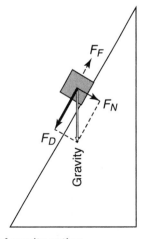

The greater the slope, the greater the force of gravity acting
in a downhill direction (F_D).

The angle of the steepest slope at which any specific type of sediment remains stable is called its **angle of repose**. The size, shape, and density of sediment particles affect their angle of repose. Water acts as a lubricant and can also affect the sediment's angle of repose. Mass movement occurs when the sediment's angle of repose is exceeded and it moves downhill. The type of mass movement that occurs depends on the slope of the land, the angle of repose of the sediment, and the amount of water present.

Mass movement is usually only the first step in erosion. It delivers sediments to the base of slopes. There, other agents of erosion, such as streams, can pick up the sediments and carry them farther.

GRAVITY

The force of gravity is universal. Every bit of matter in the universe is attracted to every other bit of matter. The force of attraction due to gravity between any two objects depends on the masses (amount of matter) of the objects and their distances from one another. The more mass an object has, the greater the force with which it attracts other objects. The greater the distance between two objects, the weaker the force of gravity attracting them to one another. This relationship can be expressed in a simple mathematical formula:

$$f \propto \frac{m_1 m_2}{r^2}$$

In this equation, f is the force of gravity, $\propto$ means proportional to, m_1 and m_2 are the masses of the two objects, and r is the distance between the centers of the masses.

So, what's up with the boxes on the slope? The matter in the box is attracted to the matter in Earth. The force of attraction between the two objects acts in a direction between the centers of the two objects. Thus, the force of gravity between the box and Earth is drawn from the center of the box toward Earth's center. That arrow is labeled "Gravity." Now look at the box on the gentle slope. Notice that the top, left edge of the box juts out past the bottom of the box. The matter in that part of the box is pulled downward toward Earth's center—but to a point that is on the downhill side of the bottom edge of the box. Therefore, some of gravity is pulling the box toward a point farther downhill from the box—or in a downhill direction. That arrow is labeled "F_D." Now look at the box on the steep slope. Notice that more of the box juts out beyond the bottom of the box on the downhill side. That means that more of the box's mass is being pulled to a point farther downhill from the bottom of the box. That is why the arrow labeled "F_D" is longer—more of the mass is being pulled to a point downhill from the box, so the force of gravity in a downhill direction is greater.

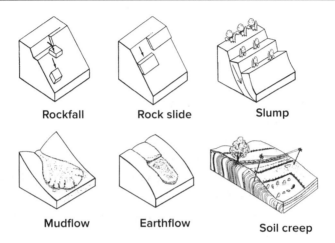

| Rockfall | Rock slide | Slump |
| Mudflow | Earthflow | Soil creep |

Rock falls occur when rock fragments formed by weathering fall from cliffs or bounce down steep slopes. **Rock slides** occur on less steep slopes when rock masses or debris slide downhill, usually triggered by earthquakes or heavy rains. A **slump** is when a huge mass of bedrock or soil slides downward from a cliff in one piece, along a curved plane of weakness, and comes to rest with its upper surface tilting toward the cliff. A **mudflow** is the rapid, downhill flow of a fluid mixture of rock, soil, and water. In an **earthflow**, a shallow layer of soil and vegetation saturated with water slowly slides downhill. **Soil creep** is the invisibly slow down-hill movement of soil, carrying the vegetation with it. Tilting of old poles, fence posts, or tombstones and bulging or broken retaining walls are all evidence of creep.

PAINLESS TIP

Check out the detailed descriptions and great photos of erosion by gravity at *http://pubs.usgs.gov/circ/1325/pdf/Sections/Section1.pdf*

BRAIN TICKLERS   Set # 2

In questions 1 to 5, write the letter of the term that best matches the definition.

1. The downhill movement of sediments under the direct influence of gravity

   a. Mudflow

2. The gradual downhill movement of a mass of soil and plant life

   b. Soil creep

3. A fluid mixture of rocks, soil, and water that flows rapidly downhill

   c. Slump

4. The rapid downhill movement of a huge block of rock or sediment in one piece

   d. Earth flow

5. The slow downhill movement of soil as it resettles after being disturbed

   e. Mass movement

6. In one or more complete sentences, explain why water is often responsible for triggering mass-wasting movements.

(Answers are on page 206.)

Moving water

Most of Earth's surface is covered by water. Therefore, it should be no surprise that moving water is one of the most important agents of erosion. Agents of erosion in which water set in motion by gravity moves sediment include streams, waves, and currents.

Streams

Streams are responsible for more erosion than all other agents of erosion combined. Water that falls to Earth as precipitation and does not seep into the ground or evaporate becomes runoff. As runoff flows downhill it can push and drag sediments along with it. Most of the sediment carried downhill by runoff ends up in streams.

Streams can exert quite a force. Imagine trying to paddle a raft up rapids. Now imagine a grain of sand or a pebble in the path of such a stream. Any loose sediment in the path of the flowing water is likely to be carried away.

Stream transport

The sediment picked up and carried by a stream is called its **load**. The amount and type of sediment that makes up a stream's load depends on the speed and volume of the stream. The faster water flows, the greater the force it exerts and the larger the size and greater the amount of sediments it can pick up and carry. At any speed, the greater the amount of water in a stream, the greater the load it can carry.

Streams transport different sediments in different ways. Large sediments like boulders, cobbles, and large pebbles are pushed or dragged downhill by rolling or sliding along the streambed.

PAINLESS TIP

Check out this video of sediment transport along a streambed. *https://www.usgs.gov/media/videos/bedload-transport-kootenai-river-near-bonners-ferry-id-part-1*

Smaller pebbles and coarse sand are light enough that they may move along the streambed in a series of bounces, hops, or leaps called **saltation**. When tiny particles like fine sand, silt, and clay are picked up, the force of the flowing water pushes them back up again whenever they begin to sink, so they are carried downhill **suspended** in the water. Some sediments dissolve in water and are carried downhill in **solution**.

Water carrying sediment is like a cutting tool. Sediments carried in a stream collide with, and knock chips off, each other and the streambed. They crush and grind up smaller particles between them. The process of chipping, crushing, grinding, and wearing away rock due to the impact of sediments is called **stream abrasion**. Abrasion wears away a stream's channel and causes the sediments carried by a stream to become smaller and rounder.

Contents of a stream bed

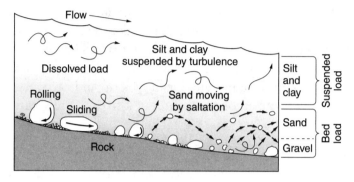

How streams transport sediment.

PAINLESS TIP

Water and wind make sediments round. Gravity and glaciers make sediments angular.

A stream's life cycle

As a stream cuts into the surface and carries away sediment, its channel gets wider, deeper, and longer. As the shape of the surface over which it flows changes, the shape and behavior of the stream changes.

Youth

Newly formed streams generally flow down steep slopes. The water flows quickly, often forming rapids. Over time, the stream wears a long, deep groove, or valley, in the surface. Stream-cut valleys typically have steep sides forming a distinctive **V-shape**.

Maturity

As a stream erodes the slope over which it flows, the slope becomes less steep. Sediment builds up at the bottom of the valley making it flat. The sides of the valley erode and the V-shape becomes wider and less steep. The slower-moving water forms curving loops called meanders. During floods, the stream overflows, spreads out over the valley floor, and drops sediments to form a broad, flat flood plain next to the stream.

Old Age

Eventually, the slopes around a stream are worn away almost completely, forming a flat lowland of gently rolling hills. The water in the stream flows slowly because the slope is so flat. Meanders become highly curved and winding and may actually loop over one another. During floods, water may gush over the land between meanders forming a **cutoff**. If the new channel cuts deep enough, part of the meander may become isolated forming an **oxbow lake**. Whatever its stage of development, the water in most streams eventually reaches the oceans.

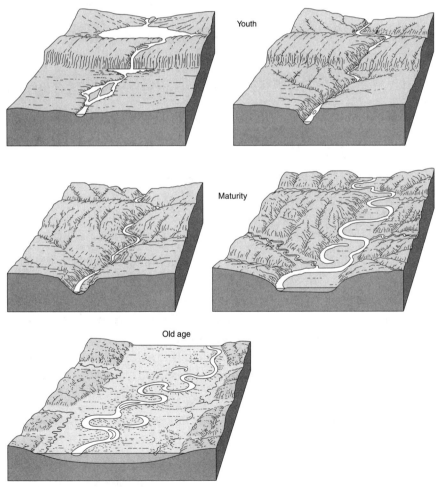

Life cycle of a stream.

Waves and currents

Ocean water is constantly in motion. Waves and currents are two examples of moving ocean water that can transport sediments and erode the land. **Waves** can be very powerful. A single wave can send tons of water crashing against a shore. The force of waves can break up rock. Loose fragments are stirred up and carried along in the turbulence of breaking waves. Wave erosion forms a number of shoreline features, some of which are shown here.

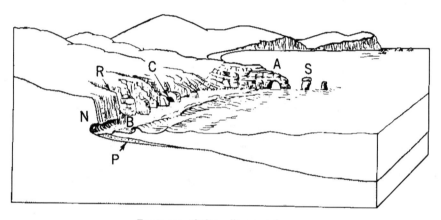

Features of shoreline erosion.
Shoreline features formed by wave erosion of a rocky coast. P—platform of abraded sediments, N—wave-cut notch at the base of a cliff, R—crevice eroded back into the rock, B—beach of sediments eroded out of the cliff, C—sea cave, A—arch, S—sea stack.

When waves strike a shore at an angle, they are reflected and interfere with incoming waves forming a **longshore current** that moves parallel to the shore. Longshore currents can move sediments along just like a stream. Along some coasts, tides flowing into and out of narrow openings form swift currents. These, too, can move large amounts of sediments.

BRAIN TICKLERS   Set # 3

1. On Earth, which agent of erosion is responsible for moving the largest amount of material?

   a. Glaciers

   b. Running water

   c. Wind

   d. Groundwater

2. Which of the rock fragments shown below was probably carried for the greatest distance by a stream?

   a.   b.   c.   d.

3. Which graph best shows the relationship between the slope of a streambed and the speed of the water flowing in the stream?

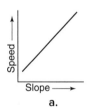 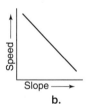  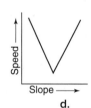

   a.   b.   c.   d.

4. As the volume of a stream increases, the amount of material that can be carried by the stream generally

   a. decreases.   b. increases.   c. remains the same.

5. Which diagram best represents a cross section of a valley that was eroded by a young stream?

   a.   b.   c.

6. The diagrams below show the same part of Earth's surface as it may have looked three million years ago and as it looks today.

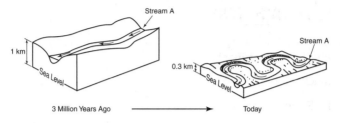

Which statement best explains the change shown?

a. The rock has been worn away.

b. The speed of the stream has increased.

c. The climate has become colder.

d. The height above sea level has become greater.

7. State three ways in which streams transport sediments.

(Answers are on page 206.)

Moving air

Wind is an important agent of erosion mainly in arid regions where there is little plant cover, and the soil, loose sediments and weathered bedrock are dry and exposed.

Wind erodes the land in two ways: deflation and abrasion. In **deflation**, the wind picks up and carries away loose particles much as a stream carries sediments. Tiny particles, like clay and dust, are carried in suspension. Larger particles, like sand, slide, roll, and bounce along the surface by saltation. In **abrasion**, exposed rock is worn away by wind-driven particles. They collide with each other and any surface rock that is exposed causing tiny chips to break off and be carried away by the wind. As a result, the exposed rock wears away and the sediments carried by wind become smaller and rounder.

Wind erosion produces a number of distinctive features. As deflation carries away layer after layer of fine sediments, a shallow depression, known as a **deflation hollow**, is formed. If the surface drops to the level of the water table, the exposed groundwater holds the soil particles together and enables plants to grow by forming an **oasis**. If the soil is made of a mixture of particle sizes, the larger ones that are too heavy to be carried away by deflation are left behind. Over time, only a continuous layer of pebbles, gravel, and other larger particles, called a **desert pavement**, remains.

THE GREAT DUST BOWL DISASTER

The Great Dust Bowl was a natural and man-made disaster that devastated the Midwestern United States in the 1930s. Encouraged by a period of abundant rainfall, people by the thousands moved west to prairie land out of reach of irrigation and plowed up the native grasses to plant row crops like corn and wheat. Then, a severe drought struck and crops began to fail. Without the deep rooted native grasses to hold the soil in place, the plowed fields were left open to wind erosion, and the soil just blew away. This created huge dust storms. Some were so thick with dust they suffocated cattle and engulfed entire towns (see the photo in Brain Tickler Set # 7, question 6, on page 204). Hundreds of thousands of families lost their farms. More than 2 million people left the Dust Bowl states of Oklahoma, Texas, Colorado, Nebraska, and Kansas during the 1930s. This created a large migrant population of poor, rural Americans searching for work right in the midst of the Great Depression.

The cause of the Great Dust Bowl disaster has a name—desertification. *Desertification* is the rapid development of deserts caused by the impact of human activities. It is mainly caused by unwise land use such as overgrazing, clearing for crops, and deforestation accelerated by natural factors such as drought. Natural cycles of drought do play a role in desertification, but without human influence, the land is less severely damaged, and it generally recovers when the drought ends. Worldwide, desertification is happening at an alarming rate. About one-third of the world's once-arable land has already been lost, and about one-third of the remaining soil is moderately to highly degraded. Only by careful soil conservation and preservation can another disaster be averted.

PAINLESS TIP

Just as a car drops lower when a tire is deflated, Earth's surface drops lower when deflation carries away sediment.

Moving ice

Glaciers are large masses of ice that form where more snow falls than melts over long periods of time. As snow piles up, its sheer weight causes the snow to compact into ice. When it becomes thick enough, the ice begins to move. Under pressure, ice behaves like a fluid and flows slowly downhill and outward due to the pull of gravity. Sediments are dragged along beneath the flowing ice and along its sides. Exposed bedrock becomes frozen onto the bottom and sides of the ice. When the glacier moves, the rock is ripped away, or **plucked**, from its original locations. Rocks torn loose as the glacier scrapes against mountains or the sides of a valley can fall on top of the glacier and get carried along as if they were on a conveyor belt.

PAINLESS TIP

Think of a glacier as a river of ice. Glacial ice flows like the water in a river, only much, much slower.

Sediments frozen onto the bottom of a glacier make it act like a huge piece of sandpaper. As a glacier moves along, it wears away exposed bedrock creating a smooth surface called **glacial polish**. Rocks frozen in the ice may also leave parallel scratches and grooves in bedrock called **glacial striations**. As a glacier flows down a V-shaped valley, it scrapes against the walls of the valley changing it to a **U-shaped valley**. See the following page.

When a glacier eventually reaches warmer regions, the ice begins to melt. If the ice flows outward faster than it melts back, the glacier's leading edge **advances**. If it melts back faster than it flows outward, the leading edge **retreats**. As climates have changed, glaciers have advanced and retreated over much of Earth's surface.

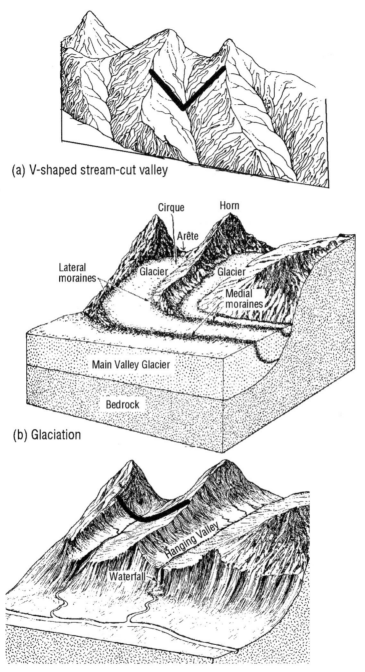

(a) V-shaped stream-cut valley

(b) Glaciation

(c) Glacier melts, leaving a U-shaped glacier-carved valley

BRAIN TICKLERS   Set # 4

1. Describe the two ways in which wind erodes Earth's surface.

2. In one or more complete sentences, explain how wind erosion forms a desert pavement.

3. Which force is mainly responsible for the movement of a glacier?

   a. Groundwater   c. Gravity

   b. Running water   d. Wind

4. A U-shaped valley scattered with large boulders was most likely formed by

   a. gravity.   c. the wind.

   b. a stream.   d. a glacier.

5. Glaciers can transport sediment by all of the following methods EXCEPT

   a. on their surface.

   b. frozen in their ice.

   c. dragged underneath or along their edges.

   d. bouncing or skipping through their ice by saltation.

(Answers are on page 206.)

Deposition

At some point, all agents of erosion begin to deposit sediment. **Deposition** is the process by which transported sediment is dropped in a new place. The same agents that erode sediment also deposit it. Several factors influence when and how deposition will take place.

Generally, as an agent of erosion slows down, it loses carrying power. Imagine you are flying a kite and the wind slows down. The wind loses some of its carrying power and the kite will begin to settle to the ground. Deposition occurs in much the same way. As an agent of erosion slows, sediment settles to the ground, or is deposited. Each agent of erosion forms distinctive landforms when it deposits sediments.

Deposition by gravity

Mass movements deposit sediment when they stop moving at the base of a slope. Most mass movement deposits are unsorted jumbles of rock fragments. Look back in the diagrams showing the different types of mass movements on page 183. Notice that each forms a different kind of deposit at the base of the slope. Rock falls and slides usually form a **talus**, or mound of broken rock that builds up at the base of a cliff or steep slope. Slump deposits have a step-like top where the block has slid down the cliff and an apron-shaped mound where the bottom of the block has spread out over the ground. Mud-flows spread out over the ground at the base of the slope in a wide, thin sheet. Huge boulders carried by the mudflow are left standing isolated on the gently sloping land. Earth flows are not as runny as mudflows. Therefore, they form a shorter, thicker deposit. Only a trained eye would recognize the series of ripples at the base of a gentle slope as a deposit formed by soil creep.

Deposition by moving water

Streams deposit sediment whenever they slow down or decrease in volume. Streams deposit sediment when they slow down because slower moving water exerts less force and therefore loses some of its carrying power. Streams deposit sediment when they decrease in volume because there is simply less water that can carry sediment. The chart below shows the sizes of the sediment particles that water can carry at different speeds.

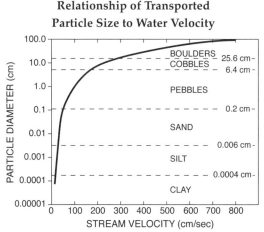

Relationship of Transported
Particle Size to Water Velocity

*This generalized graph shows the water velocity needed to maintain, but not start, movement. Variations occur due to differences in particle density and shape.

BRAIN TICKLERS   Set # 5

1. Agents of erosion generally deposit sediment when they move

   _____.

2. A deposit of jumbled and broken rock fragments that builds up at the base of a cliff or steep slope due to rock falls and slides is called a

   _____.

3. In one or two complete sentences, explain why deposits formed by Earth flows are shorter and thicker than those formed by mudflows.

4. A road is being cut through some steep, rocky hills. Why would it be a good idea for the construction crew to leave a wide, empty area on both sides of the road?

(Answers are on page 207.)

Sorting of sediments

Notice that water flowing at a speed of 150 centimeters per second can carry pebbles, sand, silt, and clay. But if the speed of the water drops below 100 centimeters per second, it cannot exert enough force to carry pebbles anymore. When that happens, the pebbles will settle out of the water. Smaller particles of sand, silt, or clay will keep going, though, until the water slows down even more. The result is a separation of particles by size in the direction in which the water is moving, or **horizontal sorting**.

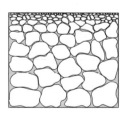

Horizontal sorting   Vertical sorting

Stream flow

Boulders   Silt   Clay   Precipitates

Cobbles

Pebbles   Sand

Horizontal and vertical sorting.

When a mixture of sediments settles through water, the sediment particles do not all settle through water at the same rate. All other factors being equal, larger sediment particles settle faster than small ones, denser particles settle faster than less dense ones, and rounded particles settle faster than flat ones. Therefore, when a mixture of sediments is deposited rapidly in a quiet body of water, the result is a layer of sediment that is **vertically sorted** according to particle size, shape, and density.

Depositional features

Deposition where water slows forms a number of distinctive stream features. When a stream curves in a meander, the water speeds up along the outside of the curve because the water is forced to cover a greater distance along the outside of the curve. But it slows down along the inside of the curve because it covers less distance along the inside of the curve. As the water slows, it drops some of its sediment load. **Bars** are mounds of sediment that usually form when water slows as it flows around the inside curve of a meander. **Oxbow lakes** form when deposition blocks off the entrance to an old meander, and it becomes separated from the stream. When a stream floods it overflows its banks. Outside of the channel, the water is shallower and moves more slowly. Floodwater begins to deposit coarser sediment as soon as it leaves the channel, forming a mound of sediment called a **levee** along the stream banks. Finer sediments are carried farther outward and are deposited forming a wide, flat **floodplain**. When a stream flows into a quiet body of water, such as an inland sea, ocean, or lake, it slows to a stop and most of the stream's sediment is deposited. The result is a **delta**, a large, flat, fan-shaped pile of sediment at the mouth of a stream.

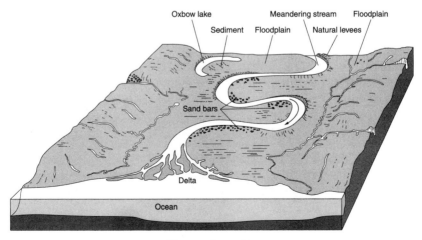

Stream deposits.

Waves and currents

Whenever waves and currents slow down, they deposit their load, building new landforms and changing the shape of shorelines. **Beaches** are formed as waves wash up against the land and erode it, then slow down, and deposit sediment along the shoreline. **Sand bars** are long, narrow piles of sand deposited in open water. Sand bars may form where waves wash beach sediments into deeper, quieter waters. They may also form where the shoreline curves away from a long-shore current. As the current curves away from the shoreline, it flows into deeper, quieter water where it slows and deposits its sediments.

BRAIN TICKLERS   Set # 6

1. At which speed would a stream be able to transport cobbles, but would not transport boulders?

   a. 50 cm/s   b. 150 cm/s   c. 250 cm/s   d. 350 cm/s

2. The separation of particles during deposition is

   a. abrasion.   c. mass movement.

   b. sorting.   d. deflation.

3. The diagram below represents a winding stream. Which pair of numbers indicates areas where deposition is most likely to occur?

   a. 1 and 4

   b. 1 and 3

   c. 1 and 2

   d. 2 and 3

4. A stream carrying a mixture of sediments flows into the ocean. Which cross section best shows the pattern of sediments deposited by the stream as it enters the ocean?

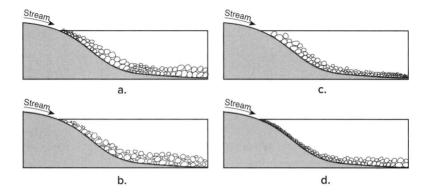

5. Quartz particles of varying sizes are dropped at the same time into deep, calm water. Which cross section best shows the settling pattern of these particles?

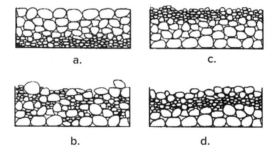

6. The map below shows the large delta that formed as the Mississippi River emptied into the Gulf of Mexico. Which process was primarily responsible for the formation of the delta?

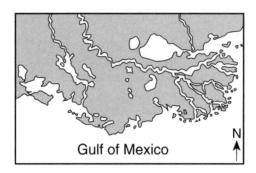

   a. Mass movement

   b. Deposition of sediment

   c. Glacial erosion

   d. Cementation of sediment

7. State three characteristics of a particle that will affect its settling rate.

(Answers are on page 207.)

Deposition by wind

When a wind slows or stops moving, the particles it is carrying settle to the ground and are deposited. **Loess** is fine, pale yellow-colored sediment deposited by winds. It is made up mainly of silt, with smaller amounts of very fine sand and clay. **Dunes** are mounds of sand deposited by wind. They are often found in barren desert regions. Dunes can also be found on large sandy beaches.

A dune often forms when windblown sand meets an obstacle such as a rock or a bush. Sand strikes the obstacle and falls to the ground. In time, the sand forms a mound that slopes gently up toward the tip of the obstacle. Wind then pushes sand grains up this slope to the crest, or top of the dune. Once over the top, the sand is deposited on the other side and the dune grows in size. As winds blow sand up one side of a dune and deposit it on the other, the entire dune moves.

Deposition by glaciers

When a glacier melts, sediments within the ice as well as those carried on top of it are released. Particles of all shapes and sizes drop to the ground in a confused jumble of unsorted sediment called **till**. The diagram here shows some of the landforms that result from deposition by glaciers.

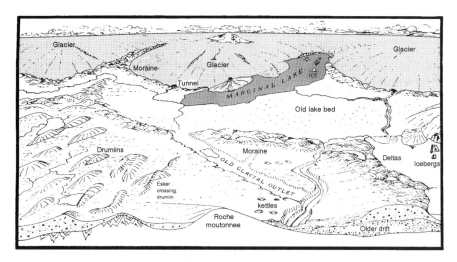

Glacial deposits.

Moraines

Deposits of till are called **moraines**. A thin, widespread layer of till that forms along the bed of the glacier is called a ground moraine. Ground moraines form gently rolling hills and valleys. Deposits of till at the melting edge of a glacier are called terminal moraines, and deposits around the side of a glacier are called lateral moraines. Terminal and lateral moraines form long, parallel ridges that mark the boundaries where a glacier once existed.

Drumlins

Sometimes a glacier retreats and then advances again over previously deposited piles of till. The moving glacier then deforms the piles of till into a long, low mound with a rounded teardrop shape called a **drumlin**. The rounded part of the drumlin points in the direction from which the glacier advanced.

Erratics

Glaciers often carry large boulders far from where they fell onto, or were frozen into, the ice. When the ice melts, the boulder is deposited on a surface composed of a different type of rock than that of the boulder. These large, isolated boulders are called **erratics**.

Outwash plains

Some glacial sediment is deposited into streams of meltwater. These sediments are then sorted by the running water and deposited in layers. Over time, these sediments build up to form a broad **outwash plain** beyond the glacier. It is called an outwash plain because the sediments it is made of were "washed out" beyond the glacier by meltwater streams.

Kettles

Sometimes a chunk of ice breaks loose from a glacier and is buried by sediment deposited by the glacier. When the ice melts, the overlying sediment sinks and forms a rounded hole called a **kettle**. If the kettle later fills up with glacial meltwater, rainwater, or groundwater, it can form a **kettle lake**.

Glaciers can also form lakes in other ways. A glacier may gouge out a large depression in Earth's surface that later fills with water. A glacial lake may also form when moraines act like a dam, blocking the flow of glacial meltwater and creating a lake.

Ice ages

Many of these glacial features formed during the last ice age. **Ice ages** are times when Earth's average temperature drops and ice covers large areas of Earth's surface. During Earth's long history there have been many ice ages. The last ice age ended about 20,000 years ago. If these landforms are found in an area, you can be sure that it was once covered by glaciers.

BRAIN TICKLERS   Set # 7

1. The diagram below shows a cross section of soil from New York State containing pebbles, sand, and clay.

   The soil was most likely deposited by

   a. an ocean current.

   b. the wind.

   c. a river.

   d. a glacier.

2. An area contains many hills made of unsorted gravel deposits, transported boulders, and small circular lakes. This area was most likely formed by

   a. stream deposition.

   b. glacial deposition.

   c. wind deposition.

   d. wave deposition.

3. The map below shows part of a drumlin field near Oswego, New York.

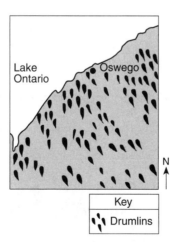

   According to the arrangement of the drumlins on the map, the glacier that formed them advanced across New York State from

   a. south to north.   **c.** northwest to southeast.

   b. east to west.   **d.** southwest to northeast.

4. The cross sections below show a three-stage sequence in the development of a glacial feature.

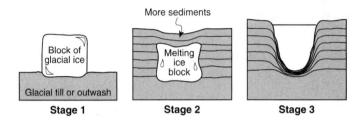

Which glacial feature has formed by the end of stage 3?

a. Kettle lake

c. Finger lake

b. Drumlin

d. Parallel scratches

5. The cross section below shows a land surface formed after two glacial advances.

A major difference between sediments in the outwash and sediments in the moraines is that the sediments deposited in the outwash are

a. larger.

c. sorted.

b. more angular.

d. older.

6. The photograph below shows farm buildings partially buried in silt.

Which erosional agent most likely piled the silt against these buildings?

a. Glacial ice   **c.** Ocean waves

b. Wind   **d.** Mass movement

(Answers are on page 207.)

Wrapping up

- Erosion is the process by which sediments are picked up and transported.

- Deposition is the process by which sediments are dropped in a new place after being transported by erosion.

- An agent of erosion is a substance set in motion by the driving force of gravity that can transport sediment. Agents of erosion include moving water (streams, waves, currents), moving air (wind), and moving ice (glaciers).

- Mass movements occur when gravity acts directly on sediments causing them to move downhill.

- Streams carry away sediment and wear away Earth's surface and form V-shaped valleys that widen with age.

- As sediment particles settle through water, they separate or become sorted. Larger sediment particles settle faster than small ones, denser particles settle faster than less dense ones, and rounded particles settle faster than flat ones.

- Deposition by streams forms land features such as bars, levees, oxbow lakes, floodplains, and deltas.

- Wind erodes Earth's surface by deflation and abrasion.

- Wind deposits are usually made up of fine particles ranging in size from clay to silt to coarse sand. Dunes form when wind-blown sediments encounter an obstacle.

- Glaciers can transport sediments on their surface, frozen in their ice, and dragged along underneath or along their edges.

- Glacial deposits may be sorted or unsorted and include features such as moraines, outwash plains, drumlins, erratic boulders, and kettle lakes.

Brain Ticklers—The Answers

Set # 1, page 180

1. Erosion
2. Deposition
3. Gravity
4. agent of erosion
5. Any three of the following: streams, glaciers, wind, waves, or currents

Set # 2, page 184

1. e   2. d   3. a   4. c   5. b

6. Water acts as a lubricant, decreasing the friction between sediment particles and between sediment particles and the underlying surface. This decreases their angle of repose and makes it more likely that the downhill pull of gravity will exceed the friction holding the sediments in place.

Set # 3, page 189

1. b   2. d   3. a   4. b   5. b   6. a

7. Any three of the following: rolling or sliding (traction), bouncing (saltation), suspension, or solution

Set # 4, page 194

1. deflation and abrasion

2. A desert pavement is a surface covered by pebbles or other large sediments. Deflation usually only removes fine sediments. Larger, heavier sediments get left behind. Eventually, a continuous layer of pebbles or other large sediments forms that blocks the wind from reaching the finer sediments underneath.

3. c

4. d

5. d

Set # 5, page 196

1. more slowly   2. talus

3. Mudflows are more watery than earthflows; therefore, mudflows spread out more and get thinner as they are deposited.

4. The road cut will expose rock to weathering that may result in rockfalls and slides. If a wide, empty area is left at the sides of the road, rock fragments will be deposited there instead of on the road.

Set # 6, page 198

1. c   2. b   3. a   4. c   5. c   6. b

7. size; shape; density

Set # 7, page 203

1. d   2. b   3. c   4. a   5. c   6. b

Chapter 8

Earthquakes and Volcanoes

Earthquakes and volcanoes might seem very different from one another, but both are the result of Earth's internal heat. Earth's internal heat causes rock to become less dense and rise toward the surface. The hot, rising rock pushes against the overlying crust causing it to break apart and move, resulting in earthquakes. Hot molten rock that rises through the breaks forms most volcanoes.

Earthquakes

An **earthquake** is a sudden trembling of the ground. Seismologists, scientists who study earthquakes, estimate that over one million earthquakes occur each year—about one every second! During most earthquakes, the shaking is so slight it is barely noticeable. Minor earthquakes feel a bit like a railway station when a train rumbles in. Major earthquakes cause violent shaking and lurching of the ground. The ground may split open, buildings may topple, pipes and power lines may be broken, and roadways may collapse.

Causes of earthquakes

Most earthquakes are caused by **faulting**, the sudden movement of rock along fractures in Earth's crust called **faults**. These movements relieve stresses that have built up due to rising hot material pushing against Earth's crust from below. When stressed, rock can bend, much as a wooden stick can bend. Eventually, though, a point is reached where the rock can bend no further without breaking. The rock snaps and great masses of rock suddenly scrape past each other along the fault. The shock of this wrenching action jolts the crust and sets an earthquake in motion.

The point where the rock breaks is called the **focus** of an earthquake. Vibrations called earthquake waves, or **seismic waves**, spread out in all directions from the focus traveling through the surrounding rock and to Earth's surface. The earthquake is first felt at the **epicenter**, a point on Earth's surface directly above the focus. See part (a) in the figure on the following page.

Seismic waves

Seismic waves can be divided into two main categories: body waves and surface waves. Body waves travel through Earth's interior; surface waves travel along Earth's surface.

Body waves

As body waves travel through Earth's interior, changes in the density or stiffness of the rock can cause body waves to bend, or refract. There are two main types of body waves: primary waves and secondary waves.

Primary (or P-) waves travel faster than any other seismic wave and are the first (or primary) waves to reach a seismograph after an earthquake. P-waves vibrate in an accordion-like motion that compresses and expands rock in the direction in which the wave travels. They are like the waves that travel down a coiled spring when one end is pushed in and pulled out. See part (b) in the figure on the facing page. P-waves can travel through solids, liquids, and gases because all three can be compressed and expanded.

Secondary (or S-) waves travel slower than P-waves and are the second waves to reach a seismograph after an earthquake. S-waves move back and forth at right angles to the direction the wave travels. S-waves are like those that form when you shake the end of a string. See part (c) in the in figure on page 211. S-waves can only travel through solids. This is because liquids and gases don't return to their original shape after being moved by an S-wave.

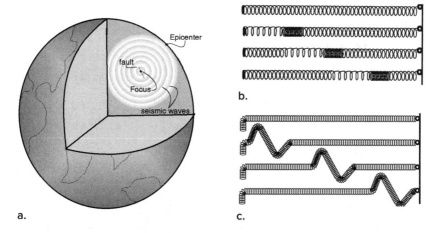

(a) Focus and epicenter; (b) P-waves; and (c) S-waves.

Surface waves

Surface waves are like water waves and travel along Earth's surface. They travel more slowly than body waves, but cause the ground to roll up and down (Love or L-waves) or twist from side to side (Rayleigh waves). For this reason, surface waves can be the most destructive type of seismic wave.

PAINLESS TIP

P is before S in the alphabet; P-waves arrive at a seismograph before S-waves. **P**-waves are **P**ush–**P**ull waves; can travel through all **P**hases (solids, liquids, gases). **S**-waves are **S**lower than P-waves; **S**econd to reach a seismograph, **S**wing back and forth, can travel through **S**olids only.

BRAIN TICKLERS   Set # 1

Write the letter of the term that best matches each definition.

1. Epicenter

2. Faulting

3. Focus

4. Primary wave

5. Secondary wave

a. The fastest moving earthquake wave

b. Earthquake waves that cannot travel through liquids

c. The point on Earth's surface directly above the focus

d. The major cause of earthquakes

e. The point beneath Earth's surface where rock breaks and moves

(Answers are on page 233.)

Detecting and measuring earthquake waves

Earthquake waves are detected and measured using an instrument called a **seismograph**. The workings of a seismograph are based on the law of inertia: objects that are at rest will tend to remain at rest. The heart of a simple seismograph is a heavy, suspended weight that tends to remain at rest while the ground moves beneath it. A pen or other recording device is attached to the weight. A recording medium such as paper is wound around a clockwork-driven drum mounted firmly in bedrock. When an earthquake causes the bedrock to move back and forth, the drum moves with it, but the heavy weight and attached pen remain motionless for a long time. As the paper moves under the motionless pen, a line is drawn on the paper that records the back and forth motion of the bedrock as shown on page 213. The line recorded on paper by a seismograph is called a **seismogram**. Most earthquake recording stations use at least three seismographs—one to measure motions in each of three dimensions, typically north–south, east–west, and up–down.

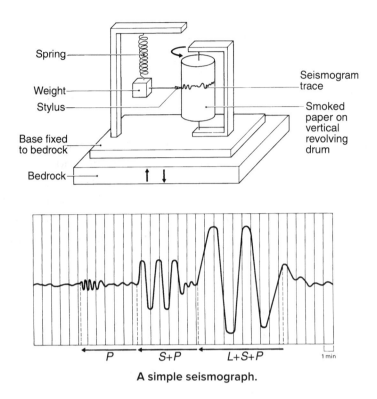

A simple seismograph.

When an earthquake occurs, the different kinds of seismic waves all start moving outward from the focus at the same time. However, since they all travel at different speeds, they do not all arrive at a seismograph at the same time. P-waves, which travel the fastest, arrive first, followed by the S-waves some time later; the surface waves arrive last. Since the speeds at which seismic waves travel is known, the difference in arrival times can be used to calculate how far away an earthquake occurred. The farther a seismograph is from the epicenter, the greater the difference between the arrival times of the P-waves and the S-waves.

By studying seismic wave data from many earthquakes, seismologists have been able to trace the paths of seismic waves as they travel through Earth. The speed of P- and S-waves changes at the boundary between regions with different properties, so they bend, or **refract**. They also travel at different speeds in substances that have different densities. This has allowed seismologists to infer that

Earth's interior is divided into four layers that differ in composition and thickness: the crust, mantle, outer core, and inner core. For
example, as seismic waves pass from the crust to the mantle, they
speed up and bend. From this evidence it is inferred that mantle rock
is denser and has a different composition than rock in the crust.

Seismologists have also discovered that between the epicenter and
an angle of 103° from the epicenter, P-waves and S-waves arrive
directly from the focus as would be expected. But, beyond 103°, there
is a band circling Earth's surface in which the S-waves *disappear*!
This is known as the S-wave **shadow zone**. Seismic stations in the
S-wave shadow zone *receive no S-waves* from an earthquake. Since
S-waves cannot travel through liquids, Earth's outer core is believed
to be a liquid.

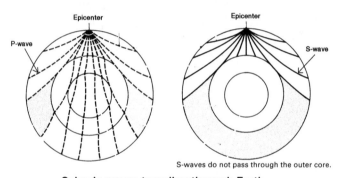

S-waves do not pass through the outer core.

Seismic waves traveling through Earth.

BRAIN TICKLERS   Set # 2

1. Scientists detect, measure, and record earthquakes using an
   instrument called a

   a. speedometer.   c. seismograph.

   b. stethoscope.   d. barometer.

2. Seismologists believe Earth's outer core is liquid because

   a. it is located in the shadow zone.

   b. P-waves speed up as they enter it.

   c. S-waves cannot travel through it.

   d. seismic waves bend when they enter it.

3. Which of the following lists the different types of seismic waves in order of the speed at which they travel, from fastest to slowest?

   a. S-waves, P-waves, L-waves

   b. L-waves, P-waves, S-waves

   c. P-waves, L-waves, S-waves

   d. P-waves, S-waves, L-waves

4. As seismic waves pass from the crust to the mantle, they

   a. speed up.   **c.** stop moving.

   b. slow down.   **d.** reverse direction.

5. Which evidence has led scientists to conclude that there are different layers within Earth's interior?

   a. analysis of earthquake wave data

   b. measurement of Earth's diameter

   c. rock samples taken from Earth's core

   d. temperatures taken within each layer

Base your answers to questions 6 through 8 on the passage below and on your knowledge of science. The passage describes some of the properties of earthquake waves.

Earthquake Waves

An earthquake occurs when pieces of Earth's crust move, producing waves of energy called seismic waves. Two types of seismic waves that travel through Earth's interior are called P-waves and S-waves. P-waves can travel through solids and liquids, but S-waves can only travel through solids.

6. Identify *one* Earth layer in which earthquakes may occur.

7. Explain how the properties of P-waves and S-waves have been used to determine that Earth's outer core is liquid.

8. Other than moving to a new area, describe *two* actions people should take in order to prepare for the possibility of strong earthquakes.

(Answers are on page 233.)

Measuring Earthquakes

As seismic waves travel farther from the focus, they lose energy and their effects lessen. The Richter and the Mercalli scales are used to measure the strength of earthquakes. The Richter scale is based on the energy released by the earthquake. It is based on direct measurements of the motions of the crust using seismic instruments. The greater the energy released by an earthquake, the greater the height of a seismic wave on a seismogram. The Richter scale consists of numbers ranging from 0 to 9. Each increase of 1 unit on the Richter scale corresponds to a tenfold increase in wave height. And every tenfold increase in wave height corresponds to a hundredfold increase in energy! Thus, the wave heights of a magnitude 2 earthquake are ten times higher than those of a magnitude 1 earthquake, and release *one hundred times as much energy.*

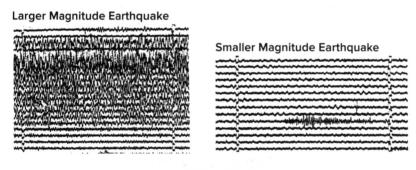

Richter Scale.

The amount of energy released by an earthquake is not the only factor affecting how much destruction it causes. Because waves lose energy as they travel through Earth, distance from the epicenter affects how much damage they will cause. Different materials react differently when shaken by earthquake waves. Therefore, a magnitude 5 earthquake may cause very different amounts of damage in two different places. Since humans are concerned about the damage to their land and property, an earthquake scale based on damage caused was developed. The modified Mercalli intensity scale on the following page is based on an earthquake's effects on structures made by humans and human activities. Engineers and city planners often use this scale to help them make decisions about building and land-use issues in earthquake-prone areas.

Effects of earthquakes

The effects of an earthquake depend mainly on two factors: the energy of the seismic waves and the type of material through which they are moving. Some of the more visible earthquake effects include surface faulting, ground shaking, ground failure, and tsunamis.

Surface faulting occurs when movement along a fault causes Earth's surface to be lifted up, lowered, or shifted sideways. This causes a shift in the position of structures and surface features.

When seismic waves pass through the ground, they cause the ground to shake. Solid rock is much less affected by ground shaking than loose, water-soaked ground. Imagine tapping the side of a bowl of Jell-O with a spoon. You can hardly see the solid bowl vibrate, but the Jell-O in the bowl wiggles and jiggles noticeably. An earthquake that might only cause solid bedrock to sway slightly can cause intense shaking in nearby wet clay or in a landfill. In ground failure, intense shaking causes loose or water-soaked ground to break up and settle, forming cracks and fissures. On hillsides this can trigger landslides, slumping, and mudflows.

Tsunamis are immense sea waves caused by earthquakes beneath the ocean floor or undersea landslides. Tsunamis may be only a few meters high, but they have very long wavelengths. They also travel much faster than ordinary ocean waves. Tsunamis have been clocked moving faster than 500 kilometers per hour with wavelengths as much as 200 kilometers. When tsunamis reach shallow water, they are pushed up into huge breaking waves that can be more than 20 meters high. The force exerted by such a huge, fast moving mass of water can cause extensive damage.

The Modified Mercalli scale.

Intensity Value	Description of Effects
I	Not felt.
II	Felt by persons at rest; felt on upper floors because of sway.
III	Felt indoors. Hanging objects swing. Feels like a passing truck.
IV	Hanging objects swing. Feels like heavy truck passing, or a jolt is felt. Standing cars rock. Windows and dishes rattle. Glasses clink. In the upper range of IV, wooden frames and walls creak.
V	Felt outdoors and direction can be estimated. Sleepers are awakened. Liquids are disturbed, some spill. Small, unstable objects upset. Doors swing open and close. Shutters and pictures move.
VI	Felt by all. Persons walk unsteadily, and many are frightened and run indoors. Windows, dishes, and glassware are broken. Furniture is moved or overturned. Pictures fall from walls. Weak plaster cracks. Trees and bushes visibly shaken or heard to rustle.
VII	Difficult to stand. Noticed by drivers of cars. Hanging objects quiver. Furniture is broken. Weak chimneys break at roof line. Fall of plaster, loose bricks, masonry. Waves on ponds and water muddied. Small slides and cave-ins of sand and gravel banks.
VIII	Steering of moving cars is affected. Partial collapse of masonry structures. Chimneys and smokestacks twist and fall. Frame houses move on foundation if not bolted down. Branches broken from trees. Wet ground and steep slopes crack.
IX	General panic. Weak masonry destroyed, stronger masonry cracks, is seriously damaged. Frame structures not bolted shift off foundations. Frames cracked. Reservoirs seriously damaged. Underground pipes broken. Conspicuous cracks in ground.
X	Masonry and frame structures destroyed along with their foundations. Some bridges destroyed. Serious damage to reservoirs, dikes, dams, and embankments. Large landslides. Water thrown out of lakes, canals, rivers, etc. Rails bent slightly.
XI	Rails bent greatly. Underground pipelines completely destroyed and out of service.
XII	Damage nearly total. Large rock masses displaced. Objects thrown into the air. Lines of sight and level are distorted.

Abridged from *The Severity of an Earthquake,* U.S. Geological Survey General Interest Publication. U.S. Government Printing Office: 1989-288-913.

PAINLESS TIP

REMEMBER this in the **MID**DLE of an earthquake:
Richter–**E**nergy–**M**agnitude–**M**ercalli–**I**ntensity–**D**amage

Earthquake zones

If the location of earthquakes is plotted on a map of Earth, an interesting pattern emerges. See the map below.

Rather than being randomly spread over Earth's surface, earthquakes occur in distinct, narrow zones. These zones include the mid-ocean ridges, the rim of the Pacific Ocean (called the Ring of Fire for its many active volcanoes), and the Mediterranean Belt. As you will discover in the next chapter, the location of earthquake epicenters reveals the location of boundaries between the plates of Earth's crust.

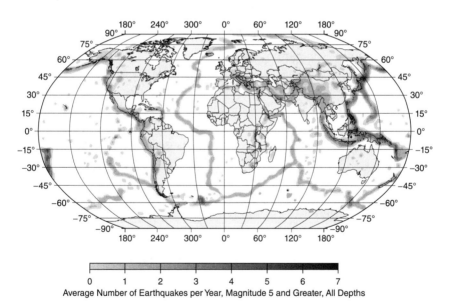

Worldwide earthquake distribution.

BRAIN TICKLERS   Set # 3

1. The greatest damage from an earthquake usually occurs

   a. far from the focus.

   b. in the shadow zone.

   c. far from the epicenter.

   d. at the epicenter.

2. The effects of an earthquake depend on all of the following, EXCEPT

   a. the types of structures in the area.

   b. air temperature at the time of the earthquake.

   c. the energy of the seismic waves.

   d. the material the ground is made of.

3. The Richter scale measures an earthquake's magnitude based on the

   a. energy of an earthquake's seismic waves.

   b. amount of damage an earthquake causes.

   c. depth at which the earthquake occurred.

   d. number of deaths and injuries caused by the earthquake.

4. Ground shaking, ground failure, surface faulting, and tsunamis are all examples of

   a. types of seismic waves.   c. causes of earthquakes.

   b. seismograph measurements.   d. effects of earthquakes.

5. Maps of the locations of an earthquake's epicenter show that most earthquakes occur

   a. in no set pattern.   c. near the equator.

   b. in narrow zones.   d. near the poles.

(Answers are on page 233.)

Volcanoes

The term **volcano** refers to both the opening in the crust through which molten rock erupts and the mountain built up by the erupted material. Volcanoes are windows into Earth's interior. Geologists can measure temperatures inside volcanoes. They can collect samples of the molten rock and gases. These observations provide evidence of Earth's internal structure and composition.

Earth's fiery interior

Beneath its cool, outer crust, Earth is glowing hot. We know this from evidence we see at the surface. Molten rock pours from volcanoes and boiling hot water and steam rise from hot springs and shoot out of geysers. Further evidence comes from mines and deep wells. As we drill deeper into Earth's crust, the temperature of the rock steadily increases—about 30°C for every 1 kilometer of depth. Considering that Earth's center is thousands of kilometers down, it is not unlikely that temperatures there exceed *6,000°C*.

Some of Earth's internal heat is left over from when it formed. Some is produced by the decay of radioactive minerals. Although Earth's interior is very hot, it is not entirely molten. In most places, the pressures within Earth are great enough to hold rock molecules together and prevent the rock from turning into a liquid. Just because the rock hasn't melted, though, doesn't mean it is hard and brittle like cold surface rocks. Much of the hot rock inside Earth is thick and pasty like modeling clay and can flow slowly under pressure.

Magma

Molten rock inside Earth is called **magma**. Magma is glowing hot, ranging in temperature from 650°C to 1,350°C. In general, as magma gets hotter it becomes more fluid. Magma also contains dissolved gases. The gases are trapped in the molten rock by the great pressures inside Earth.

A number of factors affect the melting of rock to form magma. First, different minerals melt at different temperatures. Therefore, a rock's composition affects how hot it has to be in order to melt. Second, in order to melt, the rock's temperature must be higher than its

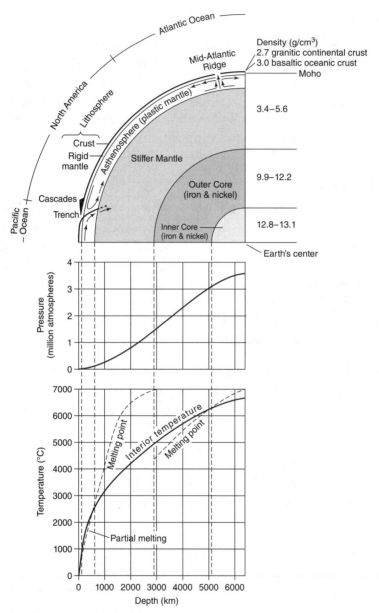

Inferred properties of Earth's interior.

melting point. Third, the pressure must be low enough not to hold the rock's molecules together and prevent melting.

As you can see in the figure above, pressures are low enough that the rock's actual temperature rises above its melting point at depths of about 70–200 kilometers. Thus, magma forms in the lower crust and upper mantle and this is the root zone of volcanoes.

From magma to lava

When solid rock melts into magma, it expands. This causes the magma to become less dense than the surrounding rock and it begins to rise. The rising magma pushes against overlying rocks and works its way upward through small cracks and fissures. This can cause the cracks and fissures to widen and can create new fractures resulting in earthquakes.

As the magma rises, the pressure on it decreases. This allows more minerals to remain molten at lower temperatures. Near the surface, magmas can melt surrounding rock forming underground pools of magma known as **magma chambers**. The molten rock in magma chambers may cool and solidify underground or work its way through cracks and fissures all the way to the surface to erupt. Magma that reaches Earth's surface is called **lava**.

Types of magma

The composition of magma depends on the type of rock from which it forms. The rocks that make up continents and ocean floors are not the same. Continental crust is mostly granite and made up of lower-density minerals with rather low melting points. Oceanic crust is mostly basalt and is made up of denser minerals with higher melting temperatures. Continental crust is also thicker than oceanic crust.

Because granite melts at a lower temperature than basalt, granitic magma can begin to form at depths of less than 100 kilometers. Basaltic magmas begin forming between depths of 100–350 kilometers where it is hotter. In both types of rock, some minerals melt at lower temperatures than others, so the magma that first forms is a slush of hot, but still solid, crystals surrounded by a thick, hot liquid. Until more than half of the melt is liquid, it behaves more like a solid than a liquid.

Basaltic magma is not only hotter than granitic magma; it is also thinner and runnier. Therefore, basaltic magma rises faster than granitic magma. It can rise so fast that dropping pressures lower the melting temperature faster than the magma is cooling. Thus, a lot of basaltic magma makes it all the way to the surface as a liquid and erupts as lava. Granitic magma is thicker and gooier, which slows the rate at which it rises. It often cools faster than dropping pressure lowers its melting temperature. Therefore, most granitic magmas solidify underground rather than reaching the surface and forming lavas.

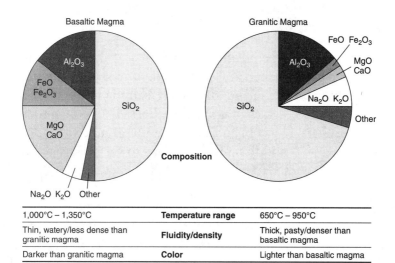

1,000°C – 1,350°C	**Temperature range**	650°C – 950°C
Thin, watery/less dense than granitic magma	**Fluidity/density**	Thick, pasty/denser than basaltic magma
Darker than granitic magma	**Color**	Lighter than basaltic magma

Types of magma.

BRAIN TICKLERS   Set # 4

1. As depth beneath Earth's surface increases, temperature and pressure

   a. increase.

   b. decrease.

   c. remain the same.

   d. decrease and then increase.

2. Magma chambers tend to form in areas beneath Earth's surface where

   a. heat builds up and pressure is reduced.

   b. pressure builds up and heat is reduced.

   c. pores in Earth are filled with groundwater.

   d. there are extensive deposits of coal, oil, or natural gas.

3. Once magma forms it tends to rise toward the surface, because melting causes the rock to

   a. contract and become denser than the surrounding rock.

   b. contract and become less dense than the surrounding rock.

   c. expand and become denser than the surrounding rock.

   d. expand and become less dense than the surrounding rock.

4. Compared to magma that forms from continental crust, magma that forms from oceanic crust is

   a. hotter, thinner, and runnier.

   b. hotter, thicker, and gooier.

   c. cooler, thinner, and runnier.

   d. cooler, thicker, and gooier.

5. As depth beneath the surface increases, the temperature at which a particular mineral will melt

   a. increases.

   b. decreases.

   c. remains the same.

   d. changes from Fahrenheit to Celsius degrees.

6. In one or two complete sentences, describe how magma differs from lava.

(Answers are on page 233.)

Volcanic eruptions

Volcanoes form where cracks lead from a magma chamber to Earth's surface. The magma in the chamber rises toward the surface through these cracks because it is less dense than the surrounding solid rocks and is under pressure. As it rises, the dissolved gases in the magma expand and are released, giving the magma an upward boost. When the magma and gases get to the surface, they burst out of the ground in a volcanic eruption.

Volcanic structures

During a volcanic eruption, volcanoes emit lava, gases, and pieces of solid rock. Erupted material spreads out in all directions around the **vent,** or central opening of a volcano. With each new eruption, more material piles up around the vent and a cone-shaped mound forms. The shape of the cone depends on the gooeyness of the magma.

Thin, runny magma tends to flow quietly out of the vent, spreading out in wide, thin sheets of lava. The result is a flat, wide cone called a **shield volcano,** named for the round shield that ancient soldiers

Main Features of a Volcano

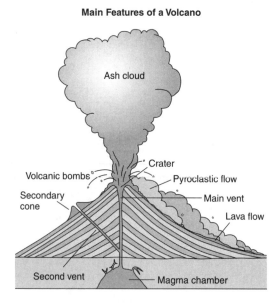

A cross section of a volcano.

used. From overhead, shield volcanoes look like a round shield lying upside down on the ground. Hawaii has a number of shield volcanoes.

In some cases, magma emerges through long, open cracks called fissures. Lava pouring out of fissures spreads out in wide, thin sheets. But instead of forming a cone as it would around a central vent, it forms a sheet that blankets the surrounding land called a **lava plateau**.

Thick, gooey lava containing a lot of dissolved gas does not flow smoothly through the vent. The vent easily becomes clogged or blocked completely. As a result gases cannot escape and pressure builds up until a violent explosion takes place. During an explosive eruption, rock and soil surrounding the vent are thrown upward along with chunks and droplets of lava blown out of the vent. The lava cools as it flies through the air and is often solid by the time it reaches the ground. Together, all of the fragments of solidified lava ejected during a volcanic eruption are called tephra. This "rain" of solid particles produces a steep, narrow cone called a **cinder cone volcano** (for the way tephra resembles cinders from a fire).

Alternating explosive and quiet eruptions form **composite cone volcanoes**—large, symmetrical cones made of alternating layers of solidified lava and volcanic rock particles.

Magma underground

Some magma moves around inside Earth but never reaches the surface. Underground flows of magma, or **igneous intrusions**, move into and through the countless underground cracks in the rocks of the crust. There, igneous intrusions may solidify into rock structures named for their shape, such as dikes, sills, laccoliths, batholiths, and stocks. **Dikes** are flat, slab-like structures that cut across layers of rock like a wall. **Sills** have a similar shape, but lie parallel to layers of rock (like a windowsill is parallel to the ground.) **Laccoliths** are large structures with a flat bottom and arched top and look like an upside-down lake. Laccolith means lake rock. Laccoliths form when magma is forced between rock layers faster than it can spread out, causing it to push the overlying layers of rock upward to form a dome mountain. Large, irregularly shaped structures are called

Types of volcanoes.

Type of Volcano	Shape	What It's Made of and How It Forms
Cinder cone		Basaltic or granitic tephra, explosive eruption ejects lava that cools as it falls through air and piles up around vent in steep-sided cone
Shield	front view / side view	Basalt lava flows; highly fluid basalt lava with low gas content pours out of vent in a quiet eruption and builds up in wide, flat layers
Composite cone	Lava flow / Ash and cinders	Layers of tephra and lava flows; steep layers of tephra from explosive eruptions interspersed with lava flows from quiet eruptions
Lava plateau		Flat layers of lava; huge amounts of lava pour out quietly from long fissures instead of central vents and flood the surrounding land with lava flow upon lava flow, forming broad plateaus

Types of Igneous Intrusions

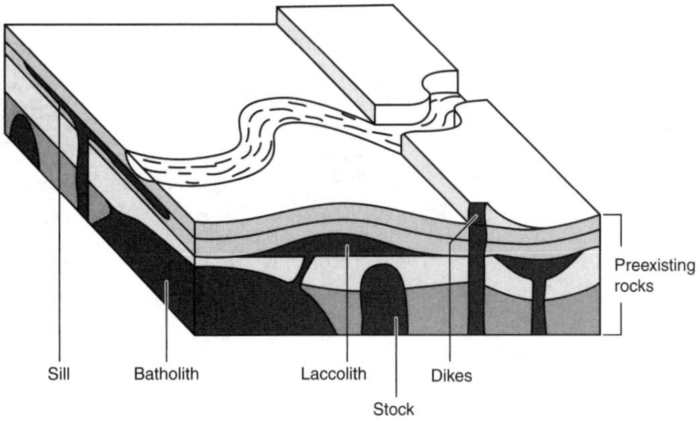

Some rock structures formed by igneous intrusions.

stocks if their area covers less than 75 square kilometers and **batholiths** if they cover more.

Zones of volcanic activity

If you plot the location of active volcanoes on a map, an interesting pattern emerges, as shown on page 230.

As you can see, active volcanoes are not evenly distributed over Earth's surface. They occur in narrow, elongated zones in some areas and are completely missing from others. For example, active volcanoes surround the Pacific Ocean, a zone known as the "Ring of Fire." However, there are almost no volcanoes along the edges of the continents bordering the Atlantic Ocean. A narrow zone of volcanoes can also be found near the centers of most oceans, producing the vast chain of undersea mountains known as the **mid-ocean ridge**. There is also a zone running along the northern edge of the Mediterranean Sea, known as the Mediterranean Belt, which includes historic volcanoes such as Mt. Etna and Mt. Vesuvius in Italy.

You may have noticed that the pattern of volcanic activity shown here and the pattern of earthquake activity shown earlier are very similar. As you will discover in the next chapter, earthquake and volcano zones mark the edges of the huge plate-like slabs of rock that make up Earth's crust.

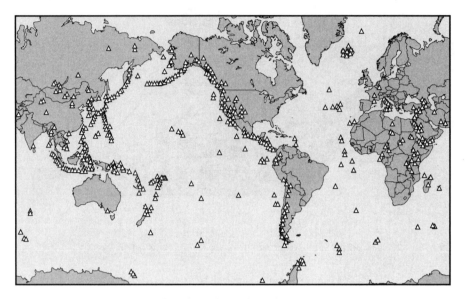

Earth's active volcanoes.

BRAIN TICKLERS   Set # 5

1. A study of rocks in an extinct volcano shows that it is composed mostly of tephra and had very steep sides. This extinct volcano was probably formed as a result of

   a. lava flows.

   b. igneous intrusions.

   c. explosive eruptions.

   d. magma solidifications.

2. Thick, pasty lava with a high gas content tends to erupt

   a. quietly.   c. from fissures in a broad sheet.

   b. explosively.   d. from undersea vents.

3. A flow of magma cools and hardens into rock before it reaches the surface. Which of the following structures most likely formed from the magma?

   a. A cinder cone   c. A geyser

   b. A lava plateau   d. A dike

4. Shield volcanoes usually have

   a. steeply sloped lava cones.   **c.** gently sloped lava cones.

   b. steeply sloped cinder cones.   **d.** gently sloped cinder cones.

5. Most of Earth's volcanoes are found

   a. evenly distributed over Earth's surface.

   b. mainly in the centers of continents.

   c. only on islands.

   d. in the same zones in which earthquakes occur.

6. Identify three types of igneous intrusions. In one or more complete sentences, explain how each intrusion is formed.

(Answers are on page 234.)

Wrapping up

- An earthquake is a sudden shaking of the ground. Its main cause is faulting.

- The effects of earthquakes include shaking and failure of the ground, surface faulting and displacement of structures, and tsunamis.

- Earthquakes are detected, measured, and recorded with seismographs. The Richter and the modified Mercalli scales are used to measure the strength of earthquakes.

- Seismic wave data provide evidence that Earth's interior is divided into four layers: the crust, mantle, outer core, and inner core.

- Earthquakes occur in narrow zones on Earth's surface.

- Temperature and pressure increase with depth beneath Earth's surface.

- A volcano is both the opening through which material erupts and the mountain built up by those materials. Magma, hot gases, and solid rock are emitted during a volcanic eruption.

- Volcanic eruptions may be explosive or quiet depending on how fluid the magma is. Quiet eruptions form shield volcanoes, explosive eruptions form cinder cone volcanoes, and a combination of quiet and explosive eruptions form composite cone volcanoes.

- Igneous intrusions are underground flows of magma that cool and solidify into rock without having reached Earth's surface. The main types of igneous intrusions include dikes, sills, laccoliths, batholiths, and stocks.

- Volcanoes tend to occur in the same narrow zones in which earthquakes occur.

Brain Ticklers—The Answers
Set # 1, page 212
1. c   2. d   3. e   4. a   5. b

Set # 2, page 214
1. c   2. c   3. d   4. a   5. a

6. crust, lithosphere, *or* mantle

7. Acceptable responses include:
   — S-waves do not travel through the liquid outer core.
   — S-waves can only travel through solids and P-waves can travel through solids and liquids.
   — by detecting which type of wave passes through the outer core.

8. Any two of the following:

   stock up on batteries/water/canned food; have an emergency-preparedness plan; change building codes/build stronger buildings; read up on what they should do in case of an earthquake/become more educated about earthquake risks; buy a radio; prepare a safety/emergency kit; conduct earthquake drills; secure furniture

Set # 3, page 220
1. d   2. b   3. a   4. d   5. b

Set # 4, page 224
1. a   2. a   3. d   4. a   5. a

6. Magma is molten rock inside Earth. Lava is molten rock that has reached Earth's surface.

Set # 5, page 230

1. c   2. b   3. d   4. c   5. d

6. Dike—an igneous intrusion that cuts across rock layers; sill—
an igneous intrusion between rock layers; laccolith—a large
intrusion with a flat bottom and arched top that looks like an
upside-down lake; stock—a large, irregularly shaped intrusion
covering less than 75 square kilometers; batholith—a large, irregularly shaped intrusion covering more than 75 square kilometers.

Chapter 9

Plate Tectonics

The more you learn about our planet Earth, the more questions come to mind. For example, after reading the last chapter, you might ask, "Why do earthquakes and volcanoes occur in narrow zones?" The theory of plate tectonics is a great example of how our understanding of Earth changes as new discoveries are made. A **theory** is a comprehensive set of ideas that explains many related observations and has been repeatedly tested. A theory that is accepted today can be revised or replaced if it no longer fits new evidence.

The Building Blocks of Plate Tectonics

The theory of plate tectonics combines a number of earlier theories such as isostasy, continental drift, and sea floor spreading into a comprehensive theory that is able to explain observations and patterns that had puzzled scientists for many years.

Isostasy

Isostasy is the theory that Earth's cool, solid crust floats on a hot, fluid mantle. By the late 1800s there was much evidence that temperature and pressure increase with depth beneath Earth's surface. Because pressure increases the density of rock, the mantle is inferred to be denser than the crust. Tests show that at high temperatures and pressures, rock behaves like a thick fluid. Because these conditions exist in the mantle, it is reasonable to infer that the mantle can act like a thick fluid. According to isostasy, the crust floats on the mantle because the mantle behaves like a thick fluid, and the crust is less dense than the mantle.

In order to understand isostasy, it is important to recall how and why things float. You probably already know that wood floats in water because it is less dense than water. The diagram below shows two blocks of wood floating in water. Both blocks are about half as dense as water, so both float exactly half in and half out of the water. Notice that the thick block floats higher above the water than the thin block. That is because half of the thick block is higher than half of the thin block. The diagram also shows same-sized blocks of wood and plastic foam floating in water. Both float in water because they are less dense than water. Notice, though, that the plastic foam floats higher than the wood. It floats higher because the foam is less dense than the wood. Thus, how high an object floats depends on both its density and its thickness.

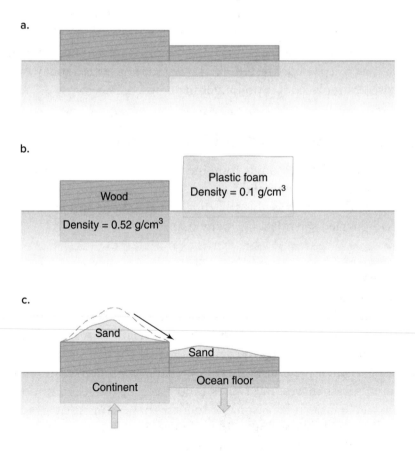

Remember, denser objects float deeper down.

Isostasy can explain why continents are higher than ocean floors. Continents are mostly granite. Ocean floors are mostly basalt. Granite is less dense than basalt. Evidence also shows that the ocean crust is thinner than continental crust. Therefore, continents float higher on the mantle than ocean floors because continental crust is thicker and less dense than ocean crust.

Isostasy can also explain vertical movements of the crust. If you pile sand on a block floating in water, it would sink a bit lower in the water. If you removed some of the sand, the block would rise and float a bit higher in the water. Now imagine what happens when the continental crust is worn away by weathering and erosion. As sediments are removed, the continental crust rises and floats a bit higher in the mantle. As these sediments are deposited on the ocean floor, it sinks a bit lower in the mantle. These up and down motions can cause rock to fracture, resulting in earthquakes. The cracks can also allow magma to reach the surface and form volcanoes.

However, isostasy cannot explain horizontal movements of the crust or inland mountain ranges. It cannot explain why earthquakes and volcanoes occur along some coasts and not others and cannot explain volcanic islands in the middle of an ocean. To explain these observations, scientists proposed other ideas.

 BRAIN TICKLERS   Set # 1

Circle the terms that best complete the statements.

1. Isostasy is the idea that the (crust / mantle) floats on the (crust / mantle).

2. Two reasons that continents float higher than ocean floors are that continents are (thicker / thinner) than ocean floors, and continents are (denser / less dense) than ocean floors.

3. If a continent is worn away by weathering and erosion, the continent will float (higher / lower) in the (crust / mantle).

4. Sediments deposited on the ocean floor will cause the ocean floor to float (higher / lower) in the (crust / mantle).

(Answers are on page 260.)

Continental drift

Map makers have long noticed how the coastlines of certain continents seem to match up. You have probably noticed how the Atlantic coastlines of Africa and South America seem to fit together like pieces of a jigsaw puzzle. In 1912, the German scientist Alfred Wegener proposed the theory of continental drift to explain matching coastlines. **Continental drift** is the theory that millions of years ago a huge landmass split apart into continents that over time have drifted apart to their present positions. Wegener and his colleagues presented striking evidence that all of the continents were once joined together in a single landmass that he called **Pangaea**, meaning "all land." The rock, mineral, and fossil evidence they gathered suggested that Pangaea broke apart into the continents about 200 million years ago.

Evidence of continental drift

Wegener borrowed the idea of blocks of crust floating on the mantle from isostasy. He cited studies of elevations and depths that showed Earth's surface is divided into two distinct levels—the continents and the ocean floors—with almost no transition between the two. He envisioned the crust as a series of side-by-side blocks floating on the mantle. Some blocks are continents; some are ocean floors.

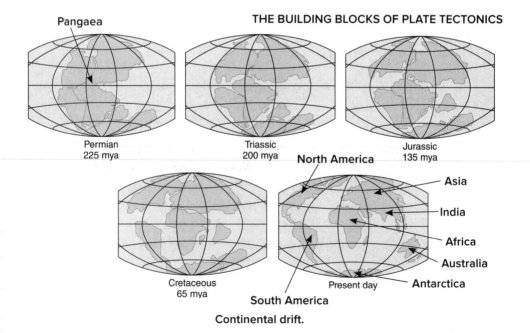

Continental drift.

Wegener pointed out *similar rock formations on the facing edges of many continents*. If the continents are put together, the rocks match almost exactly. For example, layers of rock in the Sierra Mountains near Buenos Aires closely match those of the Cape Mountains in South Africa. The diamond fields of South Africa and Brazil both occur in matching layers of a volcanic rock called kimberlite. There are similar matches of North America with Europe, Greenland with North America and Norway, Madagascar with Africa, and between other continents.

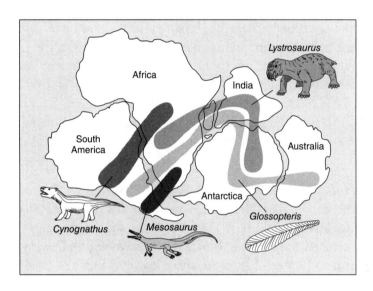

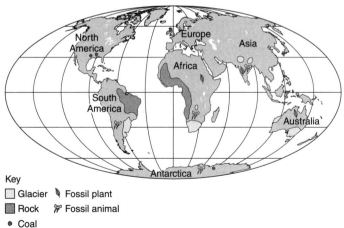

Evidence of continental drift.

Wegener also pointed to *matching fossils on different continents.* Fossils of identical land animals (*Mesosauroidae*) are found nowhere else but in southern Africa and South America. Fossils of identical trees (*Glossopteris*) are found in Australia, India, and South America. Wegener found similar matching patterns in current living things. *Lumbricus* earthworms, fresh water perch, pearl mussels, mud minnows, and garden snails found in Europe are also found only in the eastern United States. Common heather is found only in Europe and Newfoundland. Eels from Europe and eastern North America both spawn in the Sargasso Sea off the North American coast.

Coal deposits and evidence of glaciers were another source of evidence. Coal forms from the remains of plants that grow in swampy tropical forests. Yet coal deposits are found today in Antarctica and in other places with very cold climates. Much of Earth's coal was formed between 340–280 million years ago, a time called the Carboniferous period. Earth would have had to be very warm indeed for coal to form in Antarctica. But at the same time there is evidence of glaciers.

South America, southern Africa, India, and southern Australia are all warm today but show evidence of being covered by ice about 300 million years ago. If the continents were always where they are now, the ice sheet would have covered all the southern oceans and in places crossed the equator. Earth would have had to be very, very cold for this to happen. However, there is no evidence of glaciation at the same time in the Northern Hemisphere. This is hard to explain without continental drift. But if continents have drifted, then 300 million years ago they could have all been joined and some could have been located near the South Pole and some near the equator.

All of this evidence supports the idea that continents have moved over time. But scientists asked, "How can the continents move apart if the continent and ocean blocks float right up against each other?" and "What could move something as large as a continent?" Wegener could not think of a mechanism that would move the continents, so the theory of continental drift was largely ignored.

BRAIN TICKLERS   Set # 2

1. Alfred Wegener proposed the theory of

   a. continental drift.

   b. isostasy.

   c. plate tectonics.

2. All of the following ideas are part of the theory of continental drift EXCEPT

   a. all of the continents were once joined as a single, large landmass called Pangaea.

   b. about 200 million years ago, Pangaea broke apart into continents that have drifted apart ever since.

   c. the continents will keep breaking into smaller and smaller pieces until Earth's surface is covered only by small islands.

   Questions 3 and 4 refer to the map below, which shows where some continents were probably located at one time in the past and where fossils of four organisms are found today.

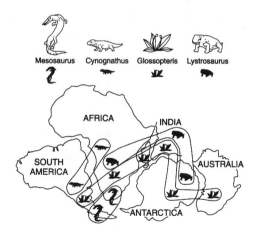

3. Which fossil shown in the diagram would be most useful for matching rocks among all of the continents shown?

4. Name two pieces of evidence shown in this diagram that suggest that these continents were once joined.

(Answers are on page 260.)

Sea floor spreading

By the 1950s oceanographic research vessels were exploring the ocean floors. New instruments revealed that the ocean floors were not flat as had been thought. Underwater mountain ranges called **mid-ocean ridges** run down the centers of most oceans. The crest of most ridges is split by deep cracks, or **rifts**. The ocean floor around the rift zones is cut by many faults. Earthquakes and volcanic activity are common in these rift zones. In other places the ocean floors plunge downward in deep, V-shaped valleys called **trenches**.

Samples taken from the ocean floors showed that most of the rocks there were younger than the rocks found on continents. In addition, the rocks on the ocean bottoms got older as you moved away from the mid-ocean ridges. This suggested that new rock was forming at the ridges and then was moving sideways away from the ridges.

Other evidence also supported this idea. Basalt, the major rock of the ocean floors, forms when lava cools and hardens. Basalt is rich in minerals containing iron. As basalt lava is cooling, crystals of these minerals line up with Earth's magnetic field. When solid, a record of the position of Earth's magnetic field is locked in the rock's crystals. Studies of ancient rocks show that the direction of Earth's magnetic field has reversed many times in the past. Research ships carrying magnetic sensors discovered that there is an identical pattern of strips

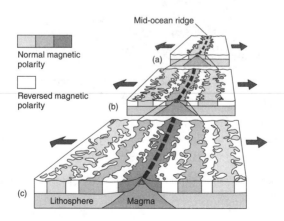

Magnetic and age patterns on ocean floor.

of normal and reversed magnetic fields on either side of the mid-ocean ridges. This also suggested that rocks were forming at the ridges and then moving sideways away from the ridges.

All of this evidence led Harry Hess of Princeton University to believe that the ocean floor was indeed spreading sideways away from the mid-ocean ridges, a theory called **sea floor spreading**. This is how he thought this was happening. Lava is extruded in the rift. The lava hardens to form new ocean floor with Earth's magnetic field "frozen" in its crystals. New lava erupts, splitting the just-formed rock and pushing it aside. The new lava hardens. As this process is repeated, new ocean floor is constantly being formed, and the floor on either side of the ridge is pushed sideways.

PAINLESS TIP

Outward = **O**lder   As you go outward from the mid-ocean ridges the bedrock gets older.

But if the ocean floor is constantly being created along the ridges, why isn't the crust getting larger? The answer was found in the trenches, the deep crevices in the ocean floor where the ocean floor bends downward sharply. Beneath the trenches, there are frequent earthquakes. Maps of these earthquakes show a pattern of increasing depth with distance from one side of the trench. Together, the plunging ocean floor and deepening pattern of earthquake depth beneath the trenches helped scientists picture what was happening. The ocean floor is plunging downward into the mantle where it melts and is destroyed in a process called **subduction**.

The rising of hot, molten material in the rifts, the sideways movement away from the rifts, and the sinking of cooler, solid crust at the trenches were strong evidence that there were convection currents in the mantle. With this final piece of evidence the picture was complete. Scientists not only had an explanation for the features and formation of the ocean floors, but also a mechanism for continental drift. Shortly thereafter, isostasy, continental drift, and ocean floor spreading were unified into the theory of plate tectonics.

New ocean floor pushes plates apart.

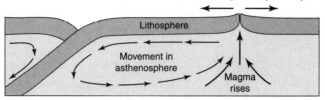

Ocean floor is created along the ridges, but is destroyed as it plunges into the mantle (forming trenches) and melts.

 BRAIN TICKLERS   Set # 3

1. According to the theory of sea floor spreading, new ocean floor is constantly being created at the

   a. rifts in the mid-ocean ridges.

   b. bottom of deep ocean trenches.

   c. centers of the continents.

   d. shorelines of oceans.

2. The diagram below shows the pattern of normal and reversed magnetic polarity and the relative age of the igneous bedrock of the ocean floor on the east side of the Mid-Atlantic Ridge. The magnetic polarity of the bedrock on the west side of the ridge has been deliberately left blank.

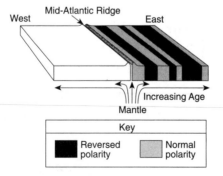

Which diagram best shows the magnetic pattern and relative age of the igneous bedrock on the west side of the ridge?

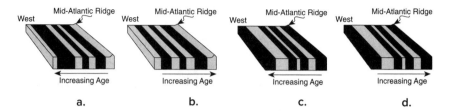

a.   b.   c.   d.

3. Ocean trenches are places where the ocean floor is

  a. buckling upward to form mountain ranges.

  b. plunging downward into the mantle forming a deep valley.

  c. not in motion in any direction.

  d. younger than at any other place on Earth.

4. Earth's crust is not increasing in size as new ocean floor is added at the mid-ocean ridges because ocean crust is being destroyed at the same rate by

  a. subduction in the trenches.

  b. weathering and erosion.

  c. earthquakes and volcanic eruptions.

  d. rip tides and deep ocean currents.

(Answers are on page 260.)

Plate Tectonics

By the 1960s, scientists had enough evidence to believe that continents and sea floors were indeed moving. The evidence showed that not only were they moving, but they were moving in many directions. This did not agree with the existing model of Earth, which thought of the continents and ocean floors as part of a single, unbroken shell of rocky crust. A solid shell cannot move in many different directions at once. Therefore, scientists had to create a new model of Earth's structure. They combined parts of isostasy, continental drift, and ocean floor spreading into a single unifying theory—**plate tectonics**. (*Tectonics* is the branch of geology dealing with the forces affecting the structure of Earth's crust.)

The lithosphere and the asthenosphere

Analysis of earthquake waves had led to the discovery that Earth's interior is made up of shells that have different properties. Based upon density differences, Earth's interior can be divided into the crust, mantle, inner core, and outer core. But if we look at the rigidity of rock instead of its density, another pattern emerges. The crust and upper mantle to a depth of about 100 kilometers is strong and rigid, forming a shell called the **lithosphere**. The region of the upper mantle between 100 and 350 kilometers in depth is weak and behaves like a thick fluid; this shell is called the **asthenosphere**.

CAUTION—Major Mistake Territory!

In plate tectonics, the lithosphere is *not just the crust*. It is made up of the crust and the uppermost part of the mantle. The crust–mantle–outer core–inner core model of Earth's interior is based mainly on composition and density. The lithosphere–asthenosphere model is based on rigidity.

Lithospheric plates

Scientists used the lithosphere/asthenosphere pattern to create a new model of Earth's structure. In this new model, the lithosphere is not an unbroken solid, but is broken up into several huge pieces

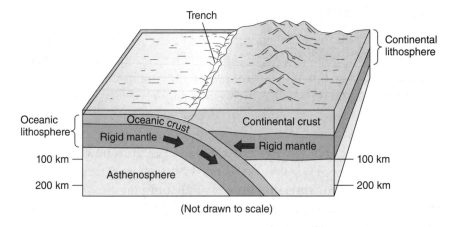

Lithosphere/asthenosphere.

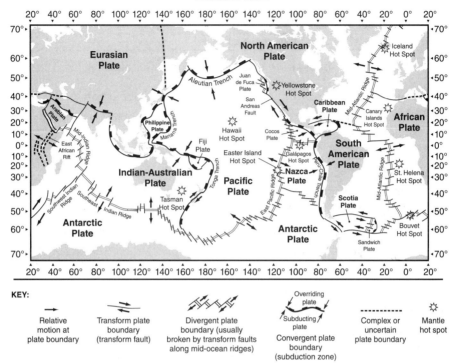

Tectonic Plates

KEY:

| Relative motion at plate boundary | Transform plate boundary (transform fault) | Divergent plate boundary (usually broken by transform faults along mid-ocean ridges) | Convergent plate boundary (subduction zone) | Complex or uncertain plate boundary | Mantle hot spot |

NOTE: Not all mantle hot spots, plates, and boundaries are shown.

or *plates*. (This is why it is called "plate" tectonics.) Earth's surface consists of about six major plates and several smaller pieces. Each plate is about 100 kilometers thick.

The plates "float" on a layer of fluid rock set in motion by huge convection currents in the mantle. The plates are carried along like bobbing corks on these fiery currents of fluid rock and collide, drift apart, or slide past one another. Continents and sea floors are moving because they are part of these plates and move along with them. It is the interaction between the edges of moving plates that explains earthquakes, volcanic activity, and crustal movements.

BRAIN TICKLERS   Set # 4

Write the letter of the term that best matches the definition.

1. A huge segment of the lithosphere   **a.** Subduction

2. A layer of soft, partially molten rock in the   **b.** Lithosphere
   upper mantle

3. A layer of strong, rigid rock made up of the   **c.** Plate
   crust and upper mantle

4. A process in which one plate is pushed   **d.** Trench
   down into the mantle beneath another

5. A deep, V-shaped valley in the ocean floor   **e.** Asthenosphere

(Answers are on page 260.)

Plate interactions

Plate boundaries are places where the edges of adjacent plates meet. As the plates move, their edges interact in one of three ways: they spread apart, they collide, or they slide past each other. Plate boundaries are regions of volcanic and earthquake activity because that is where the plates collide and scrape against each other. The interiors of plates are relatively quiet.

Divergent boundaries

Divergent boundaries are places where plates are moving apart. Divergent boundaries pull rock apart, causing it to fracture. This results in earthquakes and opens rifts through which magma can rise to the surface and solidify to form new lithosphere. The mid-ocean rifts are divergent boundaries and the mid-ocean ridges are volcanic mountains formed by magma that erupts from the rift.

Convergent boundaries

Convergent boundaries are places where plates move toward one another and collide. When plate edges collide, rocks are pushed together. This causes the rock to crumple and fold, or break and move along faults. Plates of ocean and continental crust can interact in a number of ways.

Since ocean crust is denser than continental crust, ocean crust tends to plunge under continental crust when plates collide. This forms a subduction zone where the plunging plate is forced into the mantle and melts. As one plate plunges beneath the other, the two scrape against each other producing earthquakes that get deeper as the plate slides deeper. Friction between the two plates melts rock along the boundary and produces volcanic activity directly above and parallel to the trench. This is why most trenches are bordered by volcanic island arcs or volcanic mountain chains. The collision of the two plates causes the edge of the continental plate to buckle and fold forming a long line of mountains along the edge of the continent.

If two plates carrying continental crust collide, the edges of both crumple and fold. The Andes and Himalayan Mountains formed along convergent boundaries.

Transform boundaries

Transform boundaries are places where plates slide past each other. As the plates rub and grind against each other, rock fractures form numerous faults and cause many earthquakes. If the plates separate a little, some magma may "leak" through the boundary causing small-scale volcanism. The San Andreas fault is part of a huge transform boundary along which the Pacific plate is moving past the North American plate.

Hot spots

Some volcanoes occur far from plate edges. For example, the Hawaiian volcanoes are located 4,000 kilometers from the nearest plate edge. How does plate tectonics account for volcanic activity far from plate boundaries? Canadian geophysicist John Tuzo Wilson hypothesized that there are **hot spots**—long-lasting zones of rising hot magma—at several places beneath moving plates. Large batches of magma rise from these hot spots, wedge apart cracks in the plate, and erupt, forming a volcano. As the plate moves along, older volcanoes are carried away in the direction of plate motion and new ones form over the hot spot. Thus, hot-spot volcanoes reveal the direction and rate of plate motion.

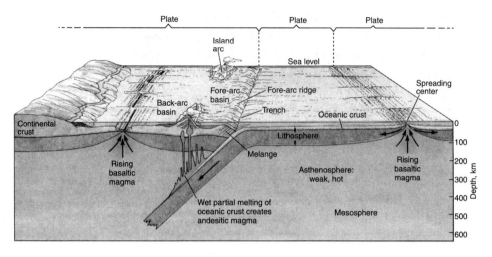

Cross section of Earth showing plate tectonic structures.

Evidence from the Hawaiian Islands supports Wilson's hot-spot hypothesis. The rocks of the Hawaiian Islands increase in age as you move away from the currently active Hawaii. The islands to the northwest are older and no longer active because the moving plate has carried them away from the hot spot. A new volcano named Loihi is forming underwater to the southeast of Hawaii. This is what would be expected if the plate is moving northwest.

The Galapagos, the Azores, and the Society Islands are also examples of volcanic islands formed by hot spots. Yellowstone National Park has hot springs, geysers, and volcanic rocks because it is located over a hot spot.

BRAIN TICKLERS   Set # 5

1. According to the theory of plate tectonics, Earth's lithosphere is

   a. a continuous, solid shell.

   b. divided into several huge, moving pieces.

   c. denser than the mantle.

   d. mostly molten.

2. The main driving force behind plate motion is

  a. Earth's rotation.

  b. heat convection in the mantle.

  c. gravitational attraction of the Sun.

  d. currents in the deep oceans.

Base your answers to questions 3–5 on the map below and the tectonic plates map on page 247. The map below shows the locations of deep-sea core drilling sites numbered 1–4. The approximate location of the East Pacific Ridge is shown by a dashed line. Point A is located on the East Pacific Ridge.

Map of Drilling Sites

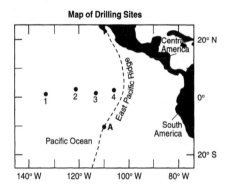

3. At point A, the East Pacific Ridge is the boundary between the

  a. Cocos Plate and the North American Plate.

  b. South American Plate and the Nazca Plate.

  c. Pacific Plate and the South American Plate.

  d. Pacific Plate and the Nazca Plate.

4. At which drilling site would the oldest igneous bedrock most likely be found?

  a. 1   **b.** 2   **c.** 3   **d.** 4

5. Compared to the thickness and density of the continental crust of South America, the oceanic crust of the Pacific floor is

  a. thinner and less dense.

  b. thinner and more dense.

  c. thicker and less dense.

  d. thicker and more dense.

Base your answers to questions 6–8 on the diagram below. The diagram shows a model of the relationship between Earth's surface and its interior.

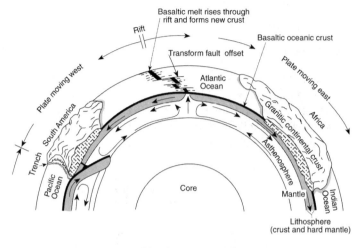

(Not drawn to scale)

6. Mid-ocean ridges (rifts) normally form where tectonic plates are

   a. converging.

   b. diverging.

   c. stationary.

   d. sliding past each other.

7. The motion of the convection currents in the mantle beneath the Atlantic Ocean appears to be mainly making this ocean basin

   a. deeper.

   b. shallower.

   c. wider

   d. narrower.

8. According to the diagram, the deep trench along the west coast of South America is caused by movement of the oceanic crust that is

   a. sinking beneath the continental crust.

   b. uplifting over the continental crust.

   c. sinking at the Mid-Atlantic Ridge.

   d. colliding with the Atlantic oceanic crust.

(Answers are on page 260.)

Effects of plate interactions

Plate interactions cause great changes in the rocks that make up the plates. They are bent, squeezed, twisted, or broken. They change position, moving up, down, or even sideways. Even the size and shape of the rocks may change. Many of these changes have produced vast landforms on Earth's surface.

You already know that forces can cause rock to break, or fracture. You may not realize that under the right conditions, they can also make rocks bend and fold. Think about a long glass rod. At room temperature it is rigid and brittle. You can bend it a tiny bit, but if you bend it too far it will break. What happens, though, if you heat the glass rod in a flame until it glows? The glass becomes pliable and can be bent. When it cools, it remains bent. In general, at high temperatures and pressures rock becomes pliable. Then, forces acting on the rock are more likely to cause the rock to bend and fold rather than breaking it. Rocks may also bend without breaking if forces are applied slowly over many years.

Folds

Under the right conditions, forces applied to rock layers can form bends or crumples that are called **folds**. Folding typically happens when rocks are forced together, or compressed. An upward arched

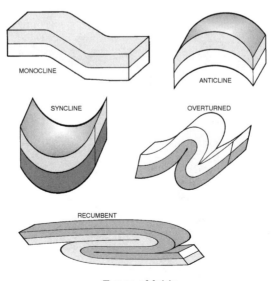

Types of folds.

fold is an **anticline**; a downward valley-like fold is a **syncline**; and the sides of a fold are called its **limbs**. Folds range from simple monoclines in which only one limb is bent to recumbent folds that bend back over themselves almost horizontally.

PAINLESS TIP

Anticlines are shaped like an A. Synclines are shaped like the top of a Y.

Joints and faults

Joints and faults are the fractures that form when rock is stressed to the point where it breaks. **Joints** are fractures along which there has not been any movement of the rock. **Faults** are fractures along which the rock has moved and one side of the fracture is displaced relative to the other side. Faulting is always associated with earthquakes. Faults are classified by the angle of the fracture and the direction in which one side moves relative to the other. Normal faults form when rock is pulled apart. Reverse and thrust faults form when rock is pushed together. Transform faults form when two parts of rock are pushed sideways past one another.

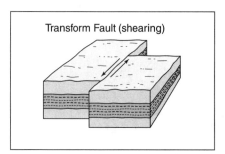

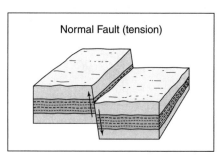

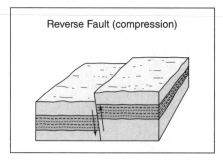

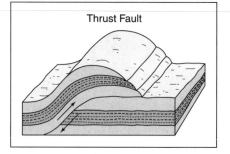

Types of faults.

PAINLESS TIP

If you stand on a fault, its **H**anging wall is over your **H**ead, and its **F**ootwall is under your **F**eet.

Remember—it's normal for your feet to slide downhill. In a normal fault, the footwall slides downward.

Mountains, plateaus, and plains

Mountains are steep-sided masses of rock that rise more than 600 meters above the surrounding land. Mountains may all seem similar at first, but mountains that form in different ways have different structures. Most mountains form at convergent or divergent plate boundaries where rocks are being pushed together or pulled apart.

Some mountains are formed when rock folds. When converging plates collide, their edges crumple and are pushed upward in folds forming fold mountains. **Fold mountains** are made mostly of sedimentary rock folded by compression. The Himalayas, Rockies, Alps, and Appalachians are ranges of fold mountains that formed along convergent plate boundaries.

Other mountains are formed by faulting due to plate movements. On one side of the fault, the rocks are forced upward. On the other side of the fault, the rocks slide downward. This creates **fault-block mountains**, mountains formed by a series of normal faults. The Sierra Nevada Mountains in California, the Grand Tetons in Wyoming, and the Wasatch Range in Utah are all examples of fault-block mountains in the United States.

Plateaus are large areas of flat land at high elevations. The rock layers beneath plateaus are usually horizontal. Plateaus are often found next to mountain ranges and are lifted up by the same forces that formed the mountains. However, the rock beneath the plateau was not folded and faulted as greatly as those of the mountains. Instead, the plateaus formed by gentle lifting of the rock layers.

Plains are large areas of flat land at low elevations. Plains are typically formed when sediments are deposited in flat layers. Most plains have also been uplifted, but not as much as plateaus. This is why plains are lower in elevation than plateaus.

a.

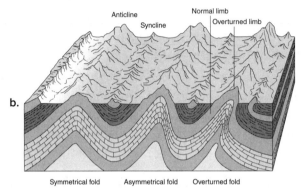

Anticline

Syncline

Normal limb

Overturned limb

b.

Symmetrical fold   Asymmetrical fold   Overturned fold

Structure of (a) fault-block mountains and (b) fold mountains.

BRAIN TICKLERS   Set # 6

1. Which block diagram best shows a transform fault?

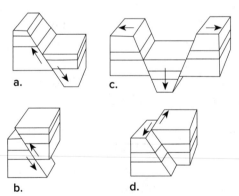

a.

c.

b.

d.

2. In the diagram below, the arrows show the direction of forces that are compressing rock layers in Earth's crust.

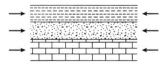

Which diagram shows the most likely result of these forces?

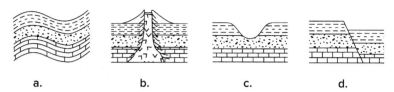

a.	b.	c.	d.

3. The diagram below shows the bedrock structure beneath a series of hills. Which process was primarily responsible for forming the hills?

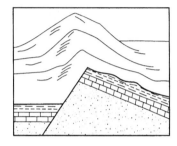

a. Folding   **b.** Deposition   **c.** Faulting   **d.** Volcanism

4. The table below contains descriptions of three different regions, A, B, and C, found in the United States.

Landscape	Bedrock	Elevation/Slopes
A	Faulted and folded layers of metamorphic rock	High elevation, steep slopes
B	Horizontal layers of sedimentary rock	Low elevation, gentle slopes
C	Thick, horizontal layers of igneous rock	Medium elevation, gentle slopes

Which list best identifies landscapes A, B, and C?

a. A–mountain, B–plain, C–plateau

b. A–plateau, B–mountain, C–plain

c. A–plain, B–plateau, C–mountain

d. A–plain, B–mountain, C–plateau

(Answers are on page 260.)

Applications of plate tectonics

Tracing plate movements has helped scientists locate valuable resources. Sometimes, when ores and oil are found on one side of an ocean, they can also be found on the other side. Plate tectonics have also helped in the search for minerals and oil in another way. Deposits of minerals and oil form only under certain conditions. By studying plate motions, geologists can predict where these conditions existed in the past or where they exist now.

Some very useful minerals form in hydrothermal deposits, places where minerals crystallize out of hot liquids. These types of deposits often form where magma heats water. Minerals in the magma, such as copper, iron, lead, zinc, and even gold, dissolve in the hot water. They are then deposited when the water cools. Plate tectonics has helped scientists find areas where the conditions needed for these deposits exist. Metal-rich sediments have been found along many plate edges. For example, metal ores have been dredged from the Indian Ocean Ridge. Sediments rich in metals have also been found along the rift in the Pacific Ocean.

Wrapping up

- A theory is a comprehensive set of ideas that explains many related observations and has been repeatedly tested.
- Isostasy is the theory that Earth's cool, solid crust floats on a hot, fluid mantle.
- Continental drift is the theory that millions of years ago a huge landmass split apart into continents that over time have drifted apart to their present positions.
- Sea floor spreading is the theory that new ocean floor is constantly being formed in the mid-ocean ridges pushing the floor on either side of the ridge sideways. In trenches, the ocean floor plunges downward into the mantle where it melts and is destroyed in a process called subduction.

- The theory of plate tectonics states that Earth's lithosphere is made up of several huge pieces or plates. These plates float on the asthenosphere, a region of the upper mantle that behaves like a thick fluid.

- Plates are set in motion by convection currents in the mantle. As the plates move, they carry the continents and sea floors along with them.

- Plate edges may separate, collide, or slide past one another. These interactions cause earthquakes, volcanic activity, and mountain building.

Brain Ticklers—The Answers

Set # 1, page 237

1. crust; mantle

2. thicker; less dense

3. higher; mantle

4. lower; mantle

Set # 2, page 241

1. a

2. c

3. Glossopteris

4. Matching shorelines; matching fossils

Set # 3, page 244

1. a   2. a   3. b   4. a

Set # 4, page 248

1. c   2. e   3. b   4. a   5. d

Set # 5, page 250

1. b	4. a	7. c
2. b	5. b	8. a
3. d	6. b	

Set # 6, page 256

1. d   2. a   3. c   4. a

Chapter 10

Earth's History

A history is a record of past events arranged in order of time. Tracing Earth's history is one of the major goals of Earth science. In order to piece together Earth's history, scientists must first determine what events happened in the past. Next they must determine the time at which those events took place. Finally, they must then arrange all of those events in order of time. A geologist's work in tracing Earth's history is a monumental task. It stretches back over the hundreds of millions of years during which life has existed on Earth to the time billions of years ago when our planet first formed.

Geologic Time

Geologic time is the entire time during which Earth has existed. Until the nineteenth century, nearly everyone believed that Earth was only a few thousand years old. Scientific evidence, though, shows that some of the rocks on Earth's surface are billions of years old. Our planet Earth is now thought to have formed about 4.5 billion years ago. Compared to this, a hundred years is a very short time. From a human viewpoint, it is hard to grasp the vastness of geologic time.

DID YOU KNOW?

If Earth's entire history was compressed into one year, the United States's entire history would fit into the last 2 seconds before midnight on December 31.

Geologic time is mostly prehistoric. That is, most of it occurred before humans kept written records of events. But that does not mean that there is no record of events that took place before humans learned to write! Earth's history is written in the materials it is made of. Every rock, fossil, and feature of Earth is evidence of a past event. Much of what is known about Earth's history, and that of life on Earth, has been learned by studying rocks.

Evidence of past events

One idea used to make sense of the evidence of past events found in rocks is uniformitarianism. It was first proposed in 1795 by the Scottish geologist James Hutton. The **theory of uniformitarianism** states that the geological processes that act on Earth today also acted on it in the past. For example, if water runs downhill today, it ran downhill a million years ago. If sediment becomes sorted as it settles through water today, it also became sorted when settling through water in the past. If basalt forms by the rapid cooling of lava today, it also formed that way in the past. Hutton's ideas can be summed up with the words "*the present is the key to the past.*"

Before Hutton, many people thought Earth's features were the result of sudden, spectacular events called catastrophes. Some thought that deep river valleys formed when the crust suddenly split apart. After long years studying rocks and natural processes, he came to the conclusion that such features could be explained in a simpler way. He thought that features such as mountains, valleys, and rock layers formed by geological processes acting over a long period of time. For example, he thought that river valleys form by the slow but never-ending erosion of the land by the river. Today, Hutton's ideas are widely accepted and used to read the story written in rocks.

For example, suppose a geologist finds that a sedimentary rock layer is made up of sorted sand grains that are rounded. Based on observations of similar layers deposited by streams today, the geologist could infer that in the past, the layer was deposited by a stream. The size of the grains could indicate how fast the water that deposited them was moving. If the rock layer also contained shells, the type of shell could show that the rock formed in a lake, not an ocean. The type of living thing that made the shells could also provide clues to the depth and temperature of the water. Microscopic pollen grains in the rock could identify plants living around the lake when the rock formed. Still other rock layers may contain evidence of glaciers, windblown deserts, volcanic eruptions, or even asteroid impacts.

This rock composed of fine grains of sand and fossils of extinct shell animals provides clues to Earth's past. Modern relatives of these shell animals live in shallow ocean waters. Therefore, geologists infer that this rock formed in shallow ocean waters.

Layers of rock that formed long ago contain the traces of events that occurred in the past. Preserved in each layer is the story of a short part of Earth's long history. By working out the age of each rock layer, the order in which they formed can be determined. Then the events they represent can be arranged in the order in which they occurred.

There are two ways to express the age of Earth materials or events. One way is to state their **absolute age**, or actual age in years. If you say that you are thirteen years old, you are stating your absolute age. Another way is to state your **relative age**, or your age as compared with another. If you say that you are older than a fourth grader and younger than your teacher, you are giving your relative age. Scientists were able to determine the relative ages of many rocks long before they could determine their absolute age. Using rocks' relative age, they were able to begin piecing together a picture of Earth's history.

BRAIN TICKLERS   Set # 1

1. Geologic time is the entire time during which

   a. Earth has existed.

   b. life has existed on Earth.

   c. the universe has existed.

   d. humans have existed.

2. Evidence suggests that the geologic processes of the past

   a. were similar to those of the present.

   b. were different from those of the present.

   c. occurred at a faster rate than those of the present.

   d. occurred at a slower rate than those of the present.

3. Which of the following describes the absolute age of a rock layer?

   a. The rock layer is older than the layer above it.

   b. The rock layer is younger than the layer below it.

   c. The rock layer is the same age as the fossils it contains.

   d. The rock layer is 1.5 million years old.

4. In one or more complete sentences, describe how you would explain the theory of uniformitarianism to a fifth grader.

(Answers are on page 284.)

Determining Relative Age

Two simple ideas used to determine the relative age of rock layers are the law of original horizontality and the principle of superposition. The **law of original horizontality** states that sediments deposited in water build up in flat, horizontal layers. It is based on countless observations of how sediments build up in layers when deposited in water. The **principle of superposition** states that in a series of layers of sedimentary rock, the bottom layer is oldest and the top layer is youngest, unless the layers have been overturned or had other layers pushed on top of them. As sediments build up over millions of years, each layer forms on top of one that was already there. Therefore, each layer is younger than the one under it and older than the one on top of it.

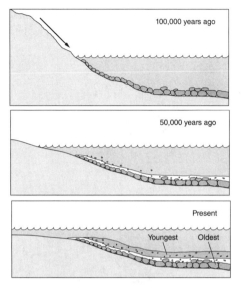

The principle of superposition.

Interpreting rock layers

However, when looking at actual rock formations, the order in which layers of rock formed is not always easy to interpret. Forces acting on the rock layers may cause them to tilt, fold, or fault. Older layers may be pushed on top of younger ones. Magma may flow through, between, or over layers of rock and harden there. Events like this disturb the sequence of the rock layers. They must be taken into account when determining the relative ages of rock layers. However, the very events that disturb rock layers also show relative age.

Joints, faults, and folds

In general, *joints, faults, or folds are younger than the rock layers in which they are found.* After all, the rock had to already be there in order to crack, fault, or fold. By working backward—unfolding or unfaulting the rock layers—the original position of the rock layers before they were disturbed can be determined. Then superposition can be used to determine their relative ages. For example, look at the folding in the first part of the diagram on page 266. If you look only at the core sample in the last picture, you might infer that layer O is the youngest and layer Y is the oldest because O is on top and Y is on the bottom. But if you look at the entire rock structure, you see that the three layers of rock have been folded and overturned. By unfolding the layers and putting them in their original position, you see that layer Y is the youngest whereas O is the oldest.

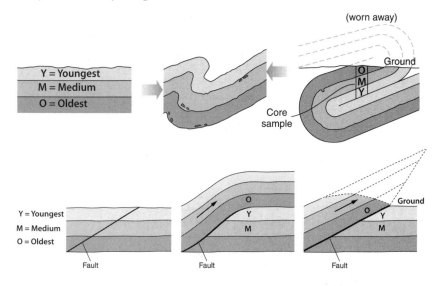

Movements of the crust may push older rock on top of younger rock.

Source: *Macmillan Earth Science*, Eric Danielson and Edward J. Denecke Jr., Macmillan Publishing Co., 1989.

The same is true of faults. Look at the second part of the diagram. Any rock layers displaced by a fault are older than the fault. But during faulting, underlying rock layers may be pushed up so that they are found on top of younger rock. Therefore, one must work backward to determine the position of the rock layers before the fault displaced them. Only then can superposition be used to determine their relative age.

Igneous intrusions and extrusions

Igneous intrusions form when magma flows into already existing rocks where it cools and hardens. Therefore, an intrusion is younger than the rock layers in which it formed. Extrusions form when lava flows out onto Earth's surface and solidifies. Extrusions are younger than any rocks beneath them, but older than any layers deposited over them. But if either is offset by a fault that crosses it, the fault is younger.

When magma or lava comes in contact with existing rock, its great heat changes the rock it touches to metamorphic rock. This is called contact metamorphism. Contact metamorphism of part of a rock layer is evidence that the rock layer is older than the igneous intrusion or extrusion. The rock layer had to already exist in order to be changed by contact with the magma or lava. On cross sections, contact metamorphism is usually shown as a series of hatch marks along the boundary where the intrusion or extrusion touches other rocks.

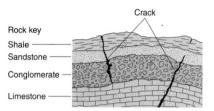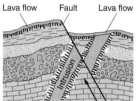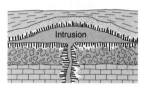

The rock layers in each of these pictures are older than the folds, joints, faults, or igneous intrusions.

Layers of rock are generally deposited in an unbroken sequence. However, if forces within Earth uplift the rocks, deposition stops and the rocks begin to erode. Erosion may wear away many layers of rock before another is laid down. This results in an **unconformity**, a break or gap in the sequence of a series of rock layers. Unconformities are like pages torn out of a book. They represent times of erosion when part of the rock record was lost. Thus, the rocks above an

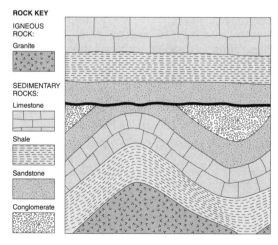

Unconformity—a gap in the rock record where the folded layers of rock were eroded before the flat layers above them were deposited.

unconformity are quite a bit younger than those below it. Still, they supply important clues to a region's history and are useful in relative dating. In cross sections, unconformities are usually shown as a wavy line to represent the uneven surfaces caused by erosion.

BRAIN TICKLERS   Set # 2

1. The diagram below shows a cross section of sedimentary rock layers that have not been overturned. Which letter most probably indicates the oldest rock layer?

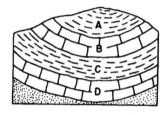

 a. A   b. B   c. C   d. D

2. Older layers of rock may be found on top of younger layers of rock as a result of

   a. weathering processes.

   b. igneous extrusions.

   c. joints in the rock layers.

   d. overturning of rock layers.

3. Unconformities (buried erosional surfaces) are good evidence that

   a. many life forms have become extinct.

   b. the earliest life forms lived in the sea.

   c. part of the geologic rock record is missing.

   d. metamorphic rocks have formed from sedimentary rocks.

4. The diagram below represents layers of rock.

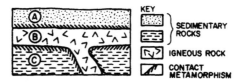

Rock layer A is inferred to be older than intrusion B because

a. layer A is composed of sedimentary rocks.

b. parts of layer A were altered by intrusion B.

c. layer B is located between layer A and layer C.

d. parts of layer C were altered by intrusion B.

Base your answers to questions 5–8 on the geologic cross section below that shows an outcrop of various types of bedrock and bedrock features in Colorado.

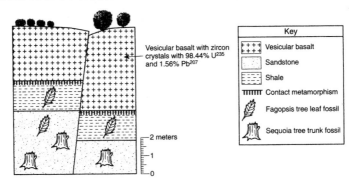

5. On the cross section, draw arrows showing the direction of movement of the rocks on both sides of the fault.

6. Use the scale in the diagram to measure how far the shale moved vertically along the fault. Express your answer to the nearest tenth of a meter.

7. In one or more complete sentences, state the evidence that supports the inference that the fault is younger than the vesicular basalt.

8. Place the geological events listed below in order by numbering them from oldest (1) to youngest (4).

_____ The fault was formed.

_____ The shale was deposited.

_____ The vesicular basalt was formed.

_____ The sandstone was deposited.

(Answers are on page 284.)

Determining Absolute Age

Suppose it is snowing when you wake up one morning and you want to know when it started to snow. One way to figure this out would be to measure the depth of the snow and the rate at which it is falling. Then you could calculate how long it took for that much snow to fall. For example, if the snow is 3 inches deep and falling at a rate of 1 inch per hour, it probably started to snow three hours ago.

Early attempts to determine absolute age were based on this type of thinking. For example, some existing sediment layers are more than 2,000 meters thick. Scientists have measured the depth of sediments deposited over objects of known age, such as shipwrecks. They have found that it takes about 15,000–30,000 years to form a layer of sediment 1 meter thick. If sediment accumulates at a rate of about 1 meter in 15,000 years, it would take 30 million years to form a 2000-meter thick layer (15,000 years/meter × 2,000 meters = 30,000,000 years). Therefore, the bottom layer would be about 30 million years old. Similar attempts were made to date rocks using weathering rates, erosion rates, and other natural processes.

PAINLESS TIP

Absolute **A**ge = **A**ctual **A**ge in years

The problem with such methods is that they are unreliable. They depend too heavily on rates that vary widely from place to place and through time. What was needed was a way to measure time by a process that does not vary. What was needed was a process that runs continuously through time and leaves a record with no gaps in it. The discovery of radioactivity provided the needed process.

Isotopes

All elements are made up of tiny bits of matter called atoms. Most elements have a number of isotopes. **Isotopes** are varieties of the same element whose atoms differ slightly in mass. For example, most carbon atoms have a mass of 12 units. This isotope is called carbon-12. But a tiny percentage of carbon atoms have a mass of 14 units. They have an extra two neutrons in their nucleus. This isotope is called carbon-14.

Radioactive decay

The most common isotopes of elements are stable, meaning their atoms do not change. But some isotopes are unstable. In a process called **radioactive decay**, the atoms of unstable isotopes break apart. During this process they give off energy and some of the small particles that make up the atom's nucleus. The end result is a stable isotope of a new element that is not radioactive. This new substance is called the **decay product**. For example, the radioactive isotope uranium-238 decays slowly into the stable decay product lead-206.

Radioactive decay takes place at a steady, constant rate. It is not affected by outside factors such as changes in temperature, changes in pressure, or chemical activity. The mineral crystals in certain types of rocks contain radioactive isotopes. The atoms of these isotopes became locked in the mineral crystals when the rock formed and have been decaying ever since. Therefore, the decay process can be used as a clock to determine the absolute age of these rocks.

Half life

The rate at which a radioactive isotope decays is known as its half-life. **Half-life** is the time it takes for one-half of the unstable radioactive isotope to change into a stable decay product. The table below lists the half-life and decay products of four common radioactive isotopes found in rocks. As you can see, some decay faster than others. If a rock contains one of these isotopes, scientists can find its absolute age.

Radioactive Decay Data.

Radioactive Isotope	Disintegration	Half-life (years)
Carbon-14	$^{14}C \rightarrow {}^{14}N$	5.7×10^3
Potassium-40	$^{40}K \begin{array}{c} \nearrow {}^{40}Ar \\ \searrow {}_{40}Ca \end{array}$	1.3×10^9
Uranium-238	$^{238}U \rightarrow {}^{206}Pb$	4.5×10^9
Rubidium-87	$^{87}Rb \rightarrow {}^{87}Sr$	4.9×10^{10}

Radioactive dating

To determine the absolute age of a rock, you compare the amount of radioisotope to the amount of decay product. For example, suppose a rock contains 50 grams of potassium-40 and 50 grams of its decay product argon-40. The entire mass of decay product was originally a radioactive isotope. Therefore, the rock originally contained 100 grams of potassium-40. It now has 50 grams of potassium-40, so exactly one-half of radioactive isotope has decayed. From the table above, we know that potassium-40 has a half-life of 1.3×109 years, or 1.3 billion years. Therefore, the rock is 1.3 billion years old.

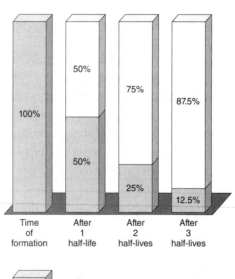

Radio isotope

Decay product

Every half-life, one half the mass of a radioactive isotope changes into its decay product.

Carbon-14 is especially useful because it can be used to date the remains of living things. All living things contain carbon, and some of that carbon is carbon-14. Carbon-14 is present in air, water, and food. As long as an organism is alive, the amount of carbon-14 in its body remains constant. Whatever decays is quickly replaced from its surroundings. However, when it dies the carbon-14 that decays is not replaced. The longer it has been dead, the less carbon-14 remains. Carbon-14 can be used to date fossils such as wood, bones, and shells. However, carbon-14 has a short half-life. After about 50,000 years the amount of carbon-14 left in a fossil is too small to be measured. Therefore, isotopes with a longer half-life must be used to date rocks and objects older than that.

Radioactive dating has enabled scientists to determine the age of Earth. Meteorites are bodies that come to Earth from outer space. Scientists believe that meteorites are made of the same material from which Earth first formed. Most burn up as they fall through the atmosphere. But some reach Earth's surface. Scientists have measured the age of these meteorites by radioactive dating. Most are about 4.5 billion years old. Rocks from the Moon have been dated at about 4.6 billion years old. Based on these measurements, it is believed that Earth is at least 4.5 to 5 billion years old.

 BRAIN TICKLERS   Set # 3

1. The absolute age of a rock is the approximate number of years ago that the rock formed. The absolute age of an igneous rock can best be determined by

   a. comparing the amount of decayed and undecayed radioactive isotopes in the rock.

   b. comparing the sizes of the crystals found in the upper and lower parts of the rock.

   c. examining the rock's relative position in a rock outcrop.

   d. examining the environment in which the rock is found.

2. Why is it unreliable to determine absolute age based on the rate at which sediments are deposited?

   a. Sediments are not radioactive.

   b. Sediments all settle through water at the same rate.

   c. Sediments may be carried far from where they first formed before being deposited.

   d. The rate at which sediments are deposited varies from place to place and over time.

3. Carbon-12 and carbon-14 are examples of

   a. two different radioactive elements.   c. compounds.

   b. isotopes of one element.   d. mixtures.

4. During radioactive decay, the atoms of a radioactive isotope

   a. disappear completely.

   b. remain unchanged.

   c. bond together in groups called molecules.

   d. break apart and form more stable atoms.

5. Why are radioactive substances useful for measuring geologic time?

   a. The disintegration of radioactive substances occurs at a predictable rate.

   b. The ratio of decay products to undecayed products remains constant in sedimentary rocks.

   c. The half-lives of most radioactive substances are shorter than 5 minutes.

   d. Measurable samples of radioactive substances are easily collected from most rock specimens.

6. A radioactive isotope commonly used to date recently formed fossils is

   a. uranium-238.   c. carbon-14.

   b. potassium-40.   d. lead-206.

7. If you started with 200 grams of a radioactive isotope, the number of grams that would be left after two half-lives is

   a. 100.   b. 75.   c. 50.   d. 25.

(Answers are on page 284.)

Fossils: Keys to the Past

Fossils are the remains, traces, or imprints of past life that have been preserved in rocks. Fossils are not actual pieces of once-living things. The actual living parts decay away, but their shape is permanently recorded in the rock as it hardens. Most fossils are preserved shapes of hard parts of living things, such as the shells, bones, or teeth of animals, or the seeds, pollen grains, or woody parts of plants. Some fossils are trace fossils. Trace fossils are not remains, but traces left by living things such as impressions of bodies, footprints, burrows, or trails left in sediment that later hardened into rock. But fossils also include such things as insects trapped in sticky tree sap that later hardened into amber, whole bodies preserved in ice or frozen soil, mummified bodies, and remains trapped in tar. **Paleontologists**, men and women who study fossils, dig up and study fossils because they are curious about life in the past.

Most fossils are found in layers of sedimentary rock. Few, if any, fossils are found in other types of rock. Igneous rocks almost never contain fossils because molten rock would burn up and destroy any remains of an organism. Metamorphic rocks rarely contain fossils because the great heat and pressure that form metamorphic rocks would damage or destroy any fossils.

How fossils form

Most fossils form when a living thing is buried soon after it dies. If the body is not buried quickly, it usually ends up in the belly of a scavenger or decays. Rapid burial is most likely to occur in water where sediments like mud, sand, and gravel are almost always settling to the bottom. This constant "shower" of sediment quickly covers any remains that sink to the bottom. Over time, the sediments harden into rock and preserve the shape of the remains as a fossil. Later, the rock may be uplifted and eroded, exposing the fossil. Since the remains of the living thing existed at the same time as the sediments in which they were buried and were fossilized as the sediments hardened into rock, the fossils in a sedimentary rock are the same age as the rock.

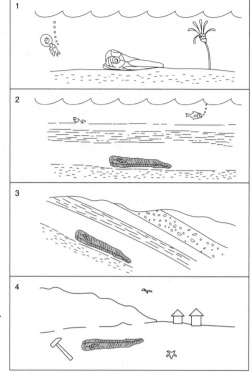

Formation of a fossil.

Index fossils

Some fossils are very useful for determining the age of the rocks in which they are found. **Index fossils** are fossils of life forms that had unique body shapes, were widespread and abundant, but only lived for a short period of time before becoming extinct. Their easily recognized body shapes and broad distribution make index fossils easy to find in rock layers worldwide. Their short existence pinpoints the time period during which those rock layers were formed.

Much of the history of life on Earth has been pieced together from fossils found in rocks. Thousands of layers of sedimentary rock that contain fossils have been found throughout the world. By working out which layers were formed earlier in time and which were formed later, scientists have also determined which fossil organisms existed earlier in time and which existed later. The fossil record shows that over time, many species have become extinct, but some species have changed, and new species have appeared.

BRAIN TICKLERS   Set # 4

1. The diagram below represents a process that occurs over a long period of time.

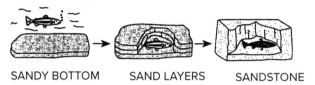

SANDY BOTTOM   SAND LAYERS   SANDSTONE

Which process is represented in this diagram?

a. Condensation   c. Photosynthesis

b. Fossilization   d. Reproduction

2. The parts of living things most often preserved as fossils are

a. soft, fleshy parts.   c. blood and other fluids.

b. hard parts.   d. hair and fur.

3. Most fossils are found in

a. igneous rocks.   c. sedimentary rocks.

b. metamorphic rocks.   d. magma chambers.

4. In order to form a fossil, the remains of a living thing must

a. decay quickly after it dies.

b. get buried soon after it dies.

c. be eaten by another living thing.

d. be trapped in molten rock.

5. Index fossils have all of the following characteristics EXCEPT

a. they are now extinct.

b. they only existed for a short period of time.

c. they are extremely rare.

d. they were abundant and widespread while alive.

6. Which conclusion can be made based on existing fossil evidence?

   a. Present life forms have always existed.

   b. Earth's environment has always been the same.

   c. Many life forms have become extinct.

   d. All life forms will remain the same in the future.

(Answers are on page 284.)

The Geologic Time Scale

Each series of layers contains a record of just a short part of Earth's history. In order to piece together a more complete picture, rocks from one place must be matched, or **correlated**, with rocks at another place of the same age. In this way, rocks in one location may fill in gaps in the record in another location. Geologists correlate rocks based on similarities in composition, color, thickness, and index fossils.

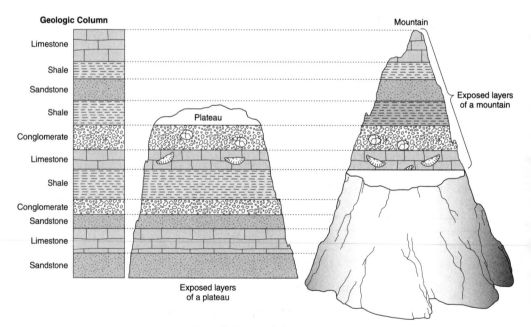

Correlating rock layers.

By the nineteenth century, geologists had correlated rocks world-wide into a single sequence called the **geologic column**. Rocks are still being added to the geologic column as more outcrops are mapped and described. Geologists then divided Earth's history into a sequence of time units called the **geologic time scale**. The geologic time scale divides time into eons, eons are subdivided into eras, eras are subdivided into periods, and periods are subdivided into epochs. At first, these time units were based only on the relative ages of fossils in the rocks. But, radioactive dating has enabled geologists to assign actual ages to the time units of the geologic time scale.

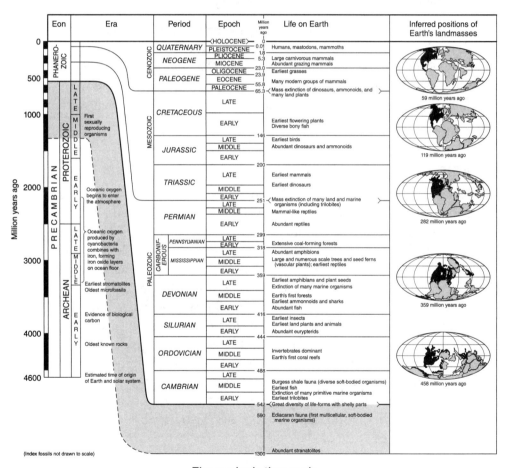

The geologic time scale.

Eons

Eons are the largest time units on the geologic time scale. The oldest is the Archean (Greek, "ancient"). Archean rocks are the oldest known on Earth and contain microscopic fossils of single-celled bacteria-like organisms. The Proterozoic (Greek, "earlier life") is the next oldest. Its rocks contain the first traces of multicelled organisms with no hard parts that could be preserved. During these eons, simple forms of life such as bacteria, microorganisms, and, later, algae and soft animals such as worms and jellyfish dominated. The most recent eon is the Phanerozoic (Greek, "visible life"), which has an abundant fossil record of preserved hard parts.

Before the traces of soft-bodied organisms and microfossils were found in Archean and Proterozoic rocks, these two eons were lumped together under the general term *Precambrian*. The oldest known rocks containing fossils of hard parts of organisms were found in a part of Britain once called Cambria. So, rocks that were older than Cambrian rocks were pre-Cambrian.

Eras

Eons are subdivided into eras based on the fossil record. Since there are so few fossils in the Archean and Proterozoic, they have not yet been formally divided into eras. The Phanerozoic is subdivided into three eras: the Paleozoic (old life), Mesozoic (middle life), and Cenozoic (recent life). These divisions are based upon the types of life forms that were dominant during those time intervals. In general, the fossil record shows that life forms have become more complex over time. During the Paleozoic era, marine invertebrates dominated and the earliest land plants and animals appeared. The Mesozoic era was dominated by reptiles, such as the dinosaurs, and saw the first mammals develop. The Cenozoic era has been dominated by mammals, including humans, and flowering plants thrived. Notice, though, that human existence has been very brief in comparison with the expanse of geologic time.

Periods

Eras are subdivided into smaller time units called periods. The names for the periods are based on the names of rock formations of that age found in many different countries. For example, the Permian is named for the province of Perm in Russia where rocks of that age are found. The Pennsylvanian and Mississippian periods are named for those places in the United States. Others are named for characteristics of the rock layers where rocks of this age were first studied, such as the Cretaceous, from Latin for "chalk."

Epochs

Epochs are the smallest unit of time on the geologic time scale. Epochs are subdivisions of periods, usually into early, middle, and late. The epochs of the Tertiary period use a jumble of Greek words describing how recent they are. For example, *Eocene*—dawn of the recent, *Miocene*—less recent, and *Holocene*—wholly recent.

BRAIN TICKLERS   Set # 5

1. The geologic time scale is divided into
   a. eons, eras, periods, and epochs.
   b. centuries, years, months, and days.
   c. mountains, valleys, plateaus, and plains.
   d. radioactive isotopes, decay products, elements, and compounds.

2. We are presently living in the
   a. Quaternary period.   c. Tertiary period.
   b. Mesozoic era.   d. Pleistocene epoch.

3. Some students plan to construct a geologic time line of Earth's history from its origin to the present time. They will use a scale of 1 meter equals 1 billion years. What should be the total length of the students' time line?
   a. 10.0 m   b. 2.5 m   c. 3.8 m   d. 4.5 m

4. A skull was discovered that has human characteristics and is about 2.8 million years old. Based on this information, during which epoch could early humans have existed?

   a. Pliocene

   b. Miocene

   c. Oligocene

   d. Eocene

5. Six hundred miles from the North Pole, researchers from the University of Rochester found the fossilized remains of *Champosaur*, a toothy, 8-foot-long extinct crocodile. The fossil was found in rocks that are 86–92 million years old. During which geologic period did *Champosaur* live near the North Pole?

   a. Ordovician

   b. Permian

   c. Triassic

   d. Cretaceous

6. Trilobites became extinct at the end of the Permian period. Which of the following sedimentary rocks could contain a trilobite fossil?

   a. 10-million-year-old granite

   b. 100-million-year-old sandstone

   c. 200-million-year-old basalt

   d. 300-million-year-old shale

(Answers are on page 284.)

Wrapping up

- Rocks contain evidence of the geologic events that make up Earth's history.
- Observations of patterns in rock layers and the location of various kinds of fossils allow inferences concerning the relative ages of rocks and the events that formed them.
- The principles of uniform process and superposition and the law of original horizontality help scientists interpret the rock record.
- The absolute age of a rock can be determined from the relative amounts of a radioactive isotope and its decay products in the rock. Half-life is the time it takes for one half of the mass of a radioactive isotope to decay.

- Fossils are the remains, traces, or imprints of past life that has been preserved in rocks. Most fossils are preserved shapes of hard parts of living things. Trace fossils include impressions of bodies, footprints, burrows, or trails left in sediment that later hardened into rock.

- Earth's rock layers have been arranged in the order in which they formed on the geologic column. On the geologic time scale, absolute ages in years have been assigned to the rocks of the geologic column.

- The geologic time scale is divided into eons, which are further subdivided into eras, periods, and epochs.

- The rock record and fossil evidence reveal that life forms and environments have changed over time.

- Scientific evidence indicates that some of the rocks on Earth's surface are several billion years old and that our planet formed about 4.5 billion years ago.

Brain Ticklers—The Answers
Set # 1, page 264

1. a   2. a   3. d

4. Answers will vary, but should convey the idea that the geologic processes that act on Earth today also acted on Earth in the past and that these processes acting over a long period of time can explain most of Earth's features.

Set # 2, page 268

1. d   2. d   3. c   4. b

5. An arrow pointing upward should be drawn on the left side of the fault and an arrow pointing downward on the right side of the fault.

6. 1.7 meters $\pm$ 0.2 meter

7. The vesicular basalt had to already exist in order to have been displaced by the fault.

8. From top of list to the bottom: 4, 2, 3, 1

Set # 3, page 273

1. a   2. d   3. b   4. d   5. a   6. c   7. c

Set # 4, page 277

1. b   2. b   3. c   4. b   5. c   6. c

Set # 5, page 281

1. a   2. a   3. d   4. a   5. d   6. d

Chapter 11

Earth in Space

Have you ever watched the Sun come up? If you look toward the east early in the morning, you can watch the first rays of sunlight stream over the horizon. This marks the beginning of the Sun's daily path across the sky. Each day, the Sun appears to move across the sky from east to west along a path that is an arc, or part of a circle.

The Sun is not the only object in the sky that appears to move across the sky. If you go out at night, you will find that the stars, Moon, and other objects seen in the sky also appear to move across the sky from east to west along a circular path. If you keep careful track of all this motion, you'll discover that almost every single one of the thousands of objects seen in the sky appears to move in the same general direction at the same speed!

DID YOU KNOW?

Objects that can be seen in the sky that are not connected with Earth or its atmosphere are known as **celestial objects**. Stars, the Sun, the Moon, the planets, and comets are all examples of celestial objects. Clouds, rainbows, halos, and other phenomena that are part of, or occur in, the atmosphere, are not.

All celestial objects appear to move across the sky from east to west along a path that is an arc, or part of a circle, at a steady 15° per hour. This equals one complete circle every day (24 hours/day × 15°/hour = 360°/day). Therefore, this motion is called **apparent daily motion**. In the Northern Hemisphere, all celestial objects move along a circular path centered very near the star *Polaris*.

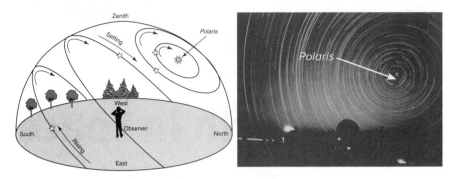

Apparent daily motion—star trails around *Polaris*

Models of the Universe

We say celestial objects *appear* to move across the sky because there are several possible reasons why an object may look like it is moving to an observer. One possibility is that the observer is standing still and the object is moving. Another is that the object is standing still and the observer is moving. Yet another is that both the observer and the object are moving, but one is moving faster, slower, or in a different direction than the other. The problem of determining which is moving—the object, the observer, or both—is not always easy to solve. Over time, people have proposed a number of different models of the universe to explain the motions of celestial objects.

The geocentric model

Early observers of the sky believed that they were standing still because their senses gave them no signs that they were moving. They *felt* as if they were standing still. Therefore, they visualized the Sun, Moon, and stars as revolving around a stationary Earth. This Earth-centered or **geocentric model** of the universe was used successfully for thousands of years to explain most observations of objects.

One effect of apparent daily motion is that the sky appears to move as if it were a single object, so early observers imagined that the sky was a single object—a huge dome. They realized that if the dome were extended far enough it would form a hollow ball, or sphere, surrounding Earth. So, they imagined a huge "sky ball," or **celestial sphere**, slowly spinning around the motionless Earth.

There were, however, some problems with this model. Celestial objects can't all be on the same sphere because some are closer and some are farther away. Early astronomers also observed that certain points of light changed position with respect to the background of stars in the sky. They called these points of light "planets," from the Greek word for "wanderer." Some of these planets even stopped, moved backward, and then moved forward again.

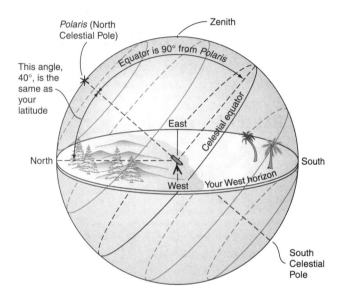

The celestial sphere.

The heliocentric model

Suppose you stand in the middle of a room and spin around, or rotate. Then, fixed objects in the room seem to move around you in circles. In 1543, Nicolaus Copernicus used this idea to suggest a model that explained the motion of celestial objects in a different way. He suggested that celestial objects seem to circle Earth because Earth rotates once a day. Stars in the Northern Hemisphere appear to circle *Polaris* because *Polaris* is aligned with Earth's axis of rotation. He also suggested that the stars and the planets (including Earth) revolved around the Sun once a year. This Sun-centered or **heliocentric model** was rejected by almost everyone because the idea that Earth was moving didn't agree with what they felt with their senses and went against their belief that Earth was the center of the universe. But, over time, the evidence for it mounted and it was eventually accepted. Our modern view of planetary motions in the solar system is based upon the heliocentric model.

PAINLESS TIP

geo = Earth   helio = Sun
centric = centered   geocentric = Earth centered
heliocentric = Sun centered

BRAIN TICKLERS   Set # 1

1. Which of the following is NOT a celestial object?

   a. The Sun   c. A rainbow

   b. The Moon   d. A star

2. In which order should the diagrams below be placed to show the position of the Sun as it appears to move across the sky in the Northern Hemisphere during one day?

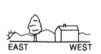

   | A | B | C | D |

   a. B → D → C → A   c. A → B → C → D

   b. A → C → B → D   d. D → B → C → A

3. The star located almost directly above our North Pole is

   a. Alpha Centauri.   c. *Polaris.*

   b. Betelgeuse.   d. Sirius.

4. An observer took a time-exposure photograph of *Polaris* and five nearby stars. How many hours were required to form these star paths?

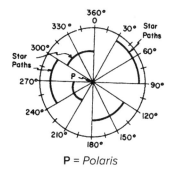

P = *Polaris*

    a. 2   b. 4   c. 6   d. 8

5. Which diagram correctly shows the heliocentric model of the motions of Earth (E) and the Moon (M) in relation to the Sun?

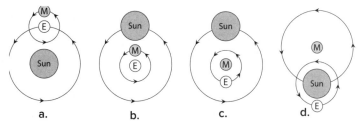

    a.   b.   c.   d.

(Answers are on page 319.)

Our Current Model of the Universe

We now know that *everything* in the universe is moving and that the universe contains a lot more than we can see with the unaided eye. Telescopes have revealed a universe filled with an almost uncountable number of stars, dust clouds, chunks of rock, and radiation.

Stars

To many people, a "star" is just a dot of light in the night sky. But a star is actually a huge sphere of hot, glowing gases. The Sun is actually an average-sized star. The Sun looks so large and bright to us because Earth is so close to it. Other stars are like the Sun, but are so far away from Earth that they just look like tiny points of light.

Stars form when gravity causes clouds of dust and gas in the universe to contract. As the dust and gas contracts, it heats up until a nuclear fusion of light elements into heavier ones occurs. Nuclear fusion in stars releases huge amounts of energy over millions of years.

Galaxies

Stars that are too dim to be seen with the unaided eye *can* be seen with a telescope. Telescopes gather more light than the human eye. The more light a telescope gathers, the larger, brighter, and more detailed the image formed. Patches of sky that look empty to the naked eye are filled with stars when viewed through a telescope. As ever-larger telescopes have been built, ever-dimmer stars have been detected, and the number of observable stars has grown from thousands to hundreds of billions.

When viewed through a telescope, some objects that look like dim smudges of light to the naked eye show up as clusters of billions of stars called **galaxies**. From what we can see, the universe contains many billions of galaxies of all sizes and shapes. Our Sun is just one of the billions of stars that make up the Milky Way Galaxy. The Sun and our solar system are located in one of the spiral arms of the galaxy about two-thirds of the way out from the center.

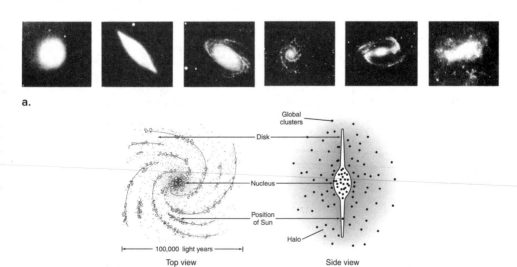

(a) Galaxies are systems containing billions of stars. Some common shapes of galaxies include spirals, ellipses, and spheres. (b) The Milky Way Galaxy.

Nebulae

Nebulae are clouds of gas and dust. Some nebulae can be seen because the gas and dust is so hot it glows or is lit up by a group of stars within the cloud. Sometimes the dust cloud is so thick that it blocks visible light, forming a black region against the background of stars.

BRAIN TICKLERS   Set # 2

1. Which statement about stars is true?

   a. They are all the same distance from Earth.

   b. They are all the same size.

   c. They shine because of the Sun's reflected light.

   d. They are great distances apart.

2. Which of the following statements best describes the difference between a galaxy and a nebula?

   a. A galaxy consists of stars; a nebula consists of dust and gas.

   b. There are two types of nebula, but only one type of galaxy.

   c. A galaxy always emits light; a nebula never emits light.

   d. A galaxy consist of matter; a nebula consists of energy.

3. Compared to stars viewed with the unaided eye, stars viewed with telescopes appear

   a. larger and brighter.   c. smaller and brighter.

   b. larger and dimmer.   d. smaller and dimmer.

4. If stars are like the Sun, why do they appear like tiny points of light?

(Answers are on page 319.)

The Solar System

The **solar system** is made up of the Sun, the planets that orbit the Sun, the satellites of those planets, and other smaller bodies that orbit the Sun, such as asteroids and comets. See the figure on page 292.

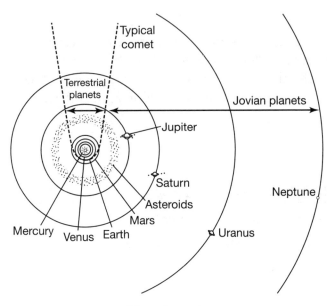

The solar system.

The Sun contains 99.9 percent of all the mass of the solar system, so the solar system pretty much is the Sun.

Formation of the solar system

The solar system formed from an interstellar cloud of gas and dust called a nebula. All matter exerts a mutual gravitational attraction on all other matter. Therefore, matter has a tendency to draw together over time, or contract. When particles of matter, such as gas molecules, draw close enough to one another, their mutual gravitational attraction may cause them to hold together, forming a larger clump of matter. This larger clump of matter exerts a greater gravitational attraction than nearby individual gas molecules because the clump has more mass. Therefore, the clump of matter more strongly attracts nearby gas molecules to itself and holds them to itself, causing the clump to become larger.

Over time, the result was further contraction and the formation of larger and larger clumps of molecules. Eventually, the contraction of the nebula (a gas cloud of molecules) formed a central body called a protostar that was surrounded by a disk of dust and gas (solar nebula).

Turbulence in the spinning disk caused it to sort itself out into concentric rings according to the mass and the speed of the material
revolving in it. Collisions between clumps of particles in the rings
resulted in them sticking together to form grains—a process called
accretion. Slowly, accretion formed larger and larger bodies called
planetesimals. As the planetesimals collided they collected into
planets, which continued to draw in dust and gas. The solar system
reached its present form when the solar nebula dissipated as the
planets literally ran out of gas and dust to draw upon.

The Sun

The **Sun** is an average-sized star, meaning some stars in the universe
are larger than the Sun and some are smaller. But even our average-
sized Sun is still more than a million times greater in volume than
Earth. Scientists estimate that the Sun contains enough matter to
sustain nuclear fusion and keep shining for 15 billion years.

The Sun is millions of times closer to Earth than any other star.
Light travels at a speed of about 300,000 kilometers per second.
It takes less than 9 minutes for light from the Sun to reach Earth.
But light from the nearest star beyond the Sun takes several years
to get to Earth. It would take thousands of years to reach that star
using our fastest rockets. Now consider that the light from the most
distant stars takes *billions* of years to reach Earth. The distances between stars are huge compared to distances within the solar system.

The planets

Planets are objects that are large enough that their gravity has pulled
them into a round shape. Their gravity is also strong enough that they
have swept up all nearby objects and debris and now orbit in a clear
path around the Sun. **Moons** are satellites, or solid bodies that orbit
planets. Beginning at the center, the major planets of the solar system
are Mercury, Venus, Earth, Mars, Jupiter, Saturn, Uranus, and Neptune.

PAINLESS TIP

A good way to remember the names of the eight major planets—**M**y
Very **E**ducated **M**other **J**ust **S**erved **U**s **N**achos.

Some basic information about the bodies in the solar system is summarized in the table below. Planets orbit the Sun in elliptical orbits. Earth's elliptical orbit is nearly circular. The major planets of the solar system divide into two groups based upon their size and composition: the four innermost small, dense **terrestrial planets** and the four large, much less dense outermost **jovian planets**.

Solar System Data.

Celestial Object	Mean Distance from Sun (million km)	Period of Revolution (d = days, y = years)	Period of Rotation at Equator	Eccentricity of Orbit	Equatorial Diameter (km)	Mass (Earth = 1)	Density (g/cm³)
SUN			27 d		1,392,000	333,000.00	1.4
MERCURY	57.9	88 d	59 d	0.206	4,879	0.06	5.4
VENUS	108.2	225.7 d	243 d	0.007	12,104	0.82	5.2
EARTH	149.6	365.26 d	23 h 56 min 4 s	0.017	12,756	1.00	5.5
MARS	227.9	687 d	24 h 37 min 23 s	0.093	6,794	0.11	3.9
JUPITER	778.4	11.9 y	9 h 50 min 30 s	0.048	142,984	317.83	1.3
SATURN	1,426.7	29.5 d	10 h 14 min	0.054	120,536	95.16	0.7
URANUS	2,871.0	84.0 y	17 h 14 min	0.047	51,118	14.54	1.3
NEPTUNE	4,498.3	164.8 y	16 h	0.009	49,528	17.15	1.8
EARTH'S MOON	149.6 (0.386 from Earth)	27.3 d	27.3 d	0.055	3,476	0.01	3.3

Physical Setting/Earth Science Reference Table—2010 Edition

The terrestrial planets

The four planets closest to the Sun are Mercury, Venus, Earth, and Mars. These terrestrial, or "Earth-like," planets resemble Earth in size and rocky composition. They are also about the same density as Earth.

The jovian planets

The four planets farthest from the Sun are Jupiter, Saturn, Uranus, and Neptune. They are called "Jupiter-like," because, like Jupiter, they are all gas giants. Gas giants have thick atmospheres made mostly of gases such as water (H_2O), methane (CH_4), and ammonia (NH_4), surrounding a small rocky or liquid core. Although they have a lot more mass than the terrestrial planets, they are less dense, so they take up much more volume. Despite their large size, the gas giants rotate very rapidly, which causes a distinct equatorial bulge.

Dwarf planets

In addition to the eight major planets, several smaller "dwarf" planets orbit the Sun. **Dwarf planets** have enough gravity that they are rounded, or nearly round in shape. But they have not swept up everything near their path and may orbit in a zone that still has many other objects in it. Currently, there are three known dwarf planets—Ceres, Pluto, and Eris. All are less than half the size of the planet Mercury and at best have only a trace of an atmosphere.

Other solar system bodies

In addition to the eight planets and three dwarf planets, many smaller objects have been found orbiting the Sun, and are therefore also considered part of the solar system. All objects that orbit the Sun but do not fit the definition of a planet or dwarf planet are called **small solar system bodies**.

Asteroids

Asteroids are irregularly shaped rocky objects orbiting the Sun. They have no atmosphere and most orbit in the so-called "asteroid belt" between the planets Mars and Jupiter. But thousands orbit farther out, like the 950-kilometer-diameter asteroid Chiron that orbits between Saturn and Uranus.

Meteoroids, meteors, and meteorites

In addition to the asteroids, there are smaller chunks of matter whose orbits cross those of the planets. When one of these chunks of matter hits Earth's atmosphere at a very high speed, it vaporizes because of friction with the air. As it streaks through the air and vaporizes, it gives off light. To an observer on Earth it appears as a streak of light. The chunk of matter is a **meteoroid**; the streak of light seen in the sky is called a **meteor**. If any of the matter survives to strike the ground it is called a **meteorite**.

Comets

Comets are small, icy masses that orbit the Sun like planets, but their orbital ellipses are highly elongated. When a comet approaches the Sun, the ices heat up and turn into a cloud of gases around the comet called a **coma**. As the comet gets closer to the Sun and warms up, more gas is given off and the coma gets larger. Particles streaming out of the Sun push gases out of the coma forming a tail that points away from the Sun.

Solar system motion

The planets all move around the Sun in nearly circular orbits. They also all orbit the Sun in the same direction and in roughly the same plane. All of the planets, except Venus and Uranus, spin in the same direction, close to the plane of the Sun's equator, and most moons also spin in the same direction as their planets, close to the plane of their planets' equators.

Gravity and orbital motion

The force that keeps planets in orbit around the Sun and moons in orbit around the planets is gravity. **Gravity** is a force of attraction that exists between all particles of matter. How does gravity keep satellites moving in a curved orbit? Imagine shooting a cannonball horizontally. If there were no gravity, the cannonball would fly off in a straight line until some force stopped it. But with gravity pulling the cannonball toward Earth's center, the path of the cannonball curves toward Earth and eventually hits its surface. If you shoot the cannonball with a more powerful charge, it will travel farther before it strikes Earth. But if you use a powerful enough charge, the cannonball would travel so far that as its path curved downward due to gravity, Earth's surface would curve away due to its spherical shape, and the ball would never strike Earth's surface. Instead, gravity would cause the cannonball to fall downward at the same rate that Earth's surface curves away from it, and it would fall unendingly in a circular path around Earth—it would be in orbit.

In the same way, it is a combination of two motions that keeps the planets moving in their curved paths around the Sun. The combination of a planet's forward motion and its motion toward the Sun due to gravity results in circular motion—the planet's orbit around the Sun. This is also what keeps the Moon in orbit around Earth.

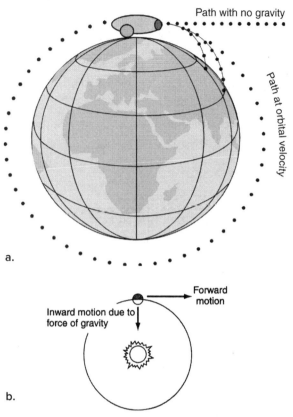

At orbital velocity an object falls toward Earth's surface in a curve that matches the curvature of Earth's surface. The orbit of a planet is the result of two motions.

BRAIN TICKLERS   Set # 3

1. Which objects are part of the solar system?

   a. Galaxies, stars, and comets

   b. The Sun, planets, and the Moon

   c. *Polaris*, planets, and galaxies

   d. Meteors, stars, and planets

2. Which object in our solar system would be classified as a star?

   a. Earth   **b.** Venus   **c.** Saturn   **d.** Sun

3. Three planets known as gas giants are
   - a. Venus, Neptune, and Jupiter.
   - c. Jupiter, Saturn, and Uranus.
   - b. Jupiter, Saturn, and Venus.
   - d. Venus, Uranus, and Jupiter.

4. Which three planets are known as terrestrial planets because of their high density and rocky composition?
   - a. Venus, Neptune, and Pluto.
   - c. Jupiter, Saturn, and Neptune.
   - b. Venus, Saturn, and Neptune.
   - d. Mercury, Mars, and Venus.

5. A meteor is best defined as
   - a. the light produced by an exploding star.
   - b. a comet-like object made of ice and dirt.
   - c. a rock fragment entering Earth's atmosphere.
   - d. a star entering Earth's atmosphere.

6. Most asteroids orbit the Sun between
   - a. Jupiter and Saturn.
   - c. Mars and Jupiter.
   - b. Uranus and Neptune.
   - d. Earth and Venus.

7. A bright object with a long tail of glowing gases is in orbit around the Sun. This object is most likely
   - a. a planet.
   - b. an asteroid.
   - c. a star.
   - d. a comet.

(Answers are on page 319.)

Earth–Moon–Sun System

Many of the events observed on Earth such as day and night, seasons, eclipses, tides, meteor showers, and comets are due to relative motion of Earth, the Moon, and the Sun.

Earth's motions

Earth, like most objects in the solar system, moves in a regular and predictable way—it rotates on its axis and revolves around the Sun.

Rotation

Rotation is a motion in which every part of an object spins around a central line called the **axis of rotation**. Earth's axis of rotation is a line passing through its center and poles. Earth rotates at a rate of 15° per hour, or one complete 360° rotation on its axis every 24 hours (15°/hour × 24 hours = 360°). Earth's axis of rotation is almost directly lined up with the star *Polaris* and is tilted at an angle of 23.5° from a perpendicular to the plane of Earth's orbit around the Sun.

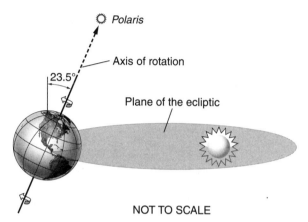

Earth rotation diagram showing plane of the ecliptic, *Polaris*, circular motion around axis of rotation, and constellations visible from the side of Earth facing away from the Sun at different times of the year.

Revolution

As Earth rotates on its axis, it also revolves around the Sun. **Revolution** is the motion of one object around another object in a path called an **orbit**. Earth completes one orbit of the Sun every 365¼ days, or one year. Earth's orbit is slightly elliptical with the Sun located at one focus of the ellipse. Therefore, Earth's distance from the Sun varies during its orbit. It is closest on January 4 and farthest away on July 4.

The plane of Earth's orbit is called the **ecliptic** because eclipses occur when Earth, the Moon, and the Sun align in this plane. Earth's axis of rotation is tilted 23.5° from a perpendicular to the ecliptic. As Earth revolves around the Sun, Earth's spin keeps its axis of rotation pointing in the same direction. In other words, Earth's axis of rotation is parallel to itself at any two points in its orbit. As a result, the Northern Hemisphere is tilted toward the Sun in June and away from the Sun in December.

As Earth revolves around the Sun, the side of Earth facing the Sun is illuminated and experiences day while the side facing away from the Sun is in darkness and experiences night. Since the stars are only visible at night, the portion of the universe whose stars are visible to an observer on Earth varies cyclically as Earth revolves around the Sun.

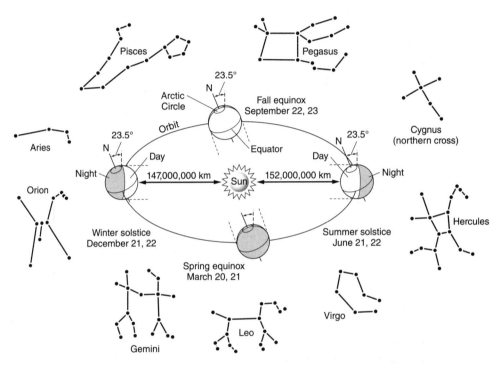

The parallelism of Earth's axis of rotation as it changes position in relation to the Sun. The shaded, nighttime side of Earth faces different constellations at different times of the year.

Effects of Earth's Motions

An Earth that rotates on an axis tilted at 23.5° as it revolves around the Sun explains a wide range of observations. In order to understand these observations, it is helpful to consider Earth's motions from two different perspectives—from space and from Earth's surface.

To represent Earth's motions from space, Earth will be shown as a sphere, and arrows will be used to show the direction of its motions.

Sunlight will be represented by arrows and the half of Earth that is in darkness will be shaded.

To represent what we see from Earth's surface, it is still useful to think of the sky as an imaginary sphere surrounding Earth. The most you would see at any one time would be half of this sphere, or a hemisphere. This hemisphere represents the sky as an observer on Earth's surface would see it. The circle formed by the intersection of the sky and the ground is called the **horizon**. The point in the sky that is right over an observer's head at any given time is called the **zenith**. An imaginary line that passes through the north and south points on the horizon and through the zenith is called the *celestial meridian*. This hemisphere model shows how the movements of celestial objects appear to an observer on Earth's surface.

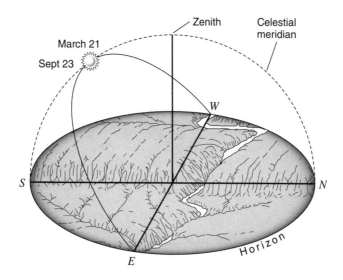

Hemisphere model of the sky as seen by an observer on Earth's surface.

The day

From space, Earth rotates on its axis from west to east or counter-clockwise when viewed from above the North Pole. Because Earth is a sphere, only one half of it is illuminated by sunlight at any given time. In one rotation, an observer carried along on Earth's surface moves from daylight into darkness and back into daylight again. This cycle of daylight and darkness is known as the **day**. The speed of Earth's rotation causes the day to be 24 hours in length.

From Earth's surface, as Earth rotates from west to east, the Sun appears to move across the sky from east to west. When a person has rotated into a position facing the Sun, the Sun appears at its highest point in the sky. This point is called solar noon.

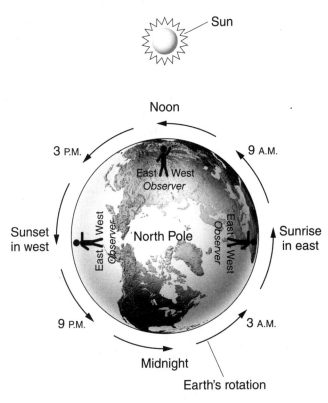

As Earth rotates from west to east, to an observer on Earth, the Sun appears to move from east to west.

The Sun is not the only celestial object that appears to move this way. Earth's rotation also causes the Moon and the stars to appear to rise in the eastern sky, arc across the sky hemisphere, and set in the western sky.

Unequal day and night

The number of hours of daylight and darkness changes as Earth's tilt relative to the Sun changes. In the diagram below, you can see that Earth's axis is tilted farthest from the Sun on December 21. Notice, though, that the boundary between daylight and darkness

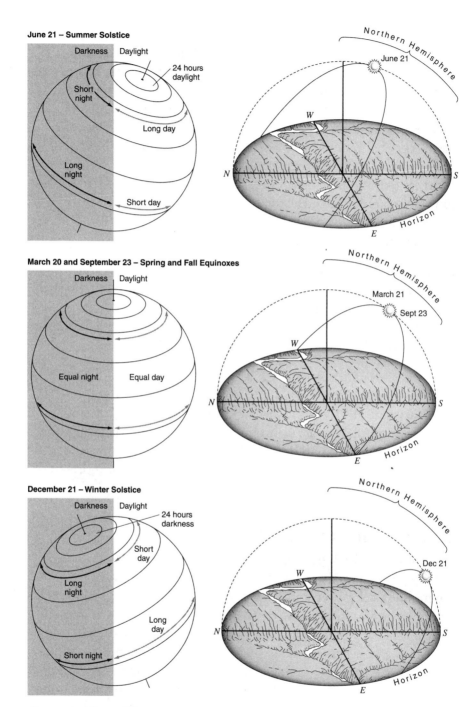

June 21 – Summer Solstice

Darkness | Daylight

24 hours daylight

Short night

Long day

Long night

Short day

Northern Hemisphere
June 21

W

N

S

E

Horizon

March 20 and September 23 – Spring and Fall Equinoxes

Darkness | Daylight

Equal night | Equal day

Northern Hemisphere

March 21

Sept 23

W

N

S

E

Horizon

December 21 – Winter Solstice

Darkness | Daylight

24 hours darkness

Short day

Long night

Long day

Short night

Northern Hemisphere

Dec 21

W

N

S

E

Horizon

Day and night at different times of the year as seen from space and from Earth's surface at about 40°N latitude.

is NOT tilted. No matter how Earth tilts, the half facing the Sun is in daylight and the half facing away is in darkness. Therefore, as Earth rotates on December 21, an observer on its surface moves in a circle around its axis of rotation that is tilted relative to the boundary between daylight and darkness. As a result, a person near the North Pole (90°N latitude) would move in a circle that never enters daylight. At the same time, a person in New York City (41°N latitude) is carried along a tilted circular path that travels mostly through darkness (~15 hours) and only travels through daylight for a short time (~9 hours). The farther south you travel on this date, the longer the daylight period, until near the South Pole, a person would never enter darkness and experience 24 hours of daylight.

On June 21, the reverse is true. Earth's axis is tilted farthest from the Sun, and a person at the South Pole would move in a circle that never enters daylight. The farther north you travel, the longer the daylight period, until at the North Pole an observer would experience 24 hours of daylight.

On March 20 and September 23, the situation is quite different. Earth is tilted neither toward nor away from the Sun, and the boundary between daylight and darkness passes through the poles. As a result, all places on Earth spend exactly one half of a rotation (12 hours) in daylight and one half of a rotation (12 hours) in darkness. To an observer on Earth's surface, the Sun appears to rise due east and set due west.

Local time and time zones

Each place on Earth experiences daylight and darkness at a different time. For example, when it is midnight in New York City, it is the middle of the afternoon in Tokyo. In the past, every place set its clocks according to when it was solar noon in that place. Clocks in New York City would be set at a different time from clocks in other cities in the same state such as Buffalo or Syracuse. Other states and countries had the same problem. To solve this problem, the countries of the world decided to set up a series of time zones. In a **time zone**, the same time is used by all places within that zone. Since Earth rotates at 15° per hour, zones were set up at 15° intervals around the globe. To avoid problems, the boundaries of time zones are shifted so that small countries, or whole states, fall within the same time zone.

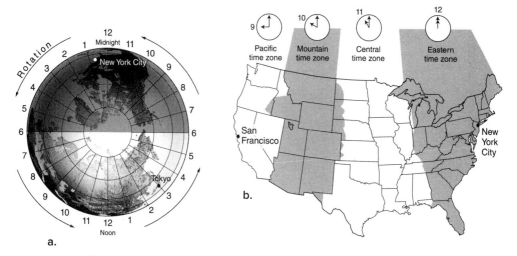

(a) Time of day at different locations. (b) Time zones of the United States.

The year

A **year** is the time it takes Earth to complete one 360° revolution around the Sun. During that time, Earth completes 365¼ rotations, so a year is 365¼ days long. Earth's revolution around the Sun causes several phenomena.

The annual traverse of the constellations

Stars are only visible at night. Therefore, the stars visible to an observer on Earth vary cyclically as Earth revolves around the Sun and its nighttime side faces different parts of the universe. Look back at the diagram on page 300.

Seasons

Sunlight does not strike all points on Earth's surface at the same angle. The two beams of sunlight approaching Earth in the diagram on the next page are identical. But notice that near the equator, the sunlight strikes Earth's surface more directly (nearer 90°) and is concentrated in a smaller area. Near the poles, the sunlight strikes Earth's surface at a lower angle (less directly) and is spread out over a larger area. The more concentrated the sunlight, the greater its heating effect. The result is higher temperatures near the equator and lower temperatures near the poles.

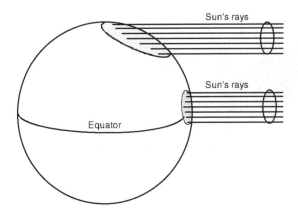

Sunlight is more direct near the equator and less direct near the poles.

Winter

Earth's tilt relative to the Sun also affects the angle at which sunlight strikes Earth's surface. On December 21, the Northern Hemisphere is tilted farthest away from the Sun. Therefore, sunlight strikes the Northern Hemisphere least directly at this time of year, and the Sun appears at its lowest altitude of the year at solar noon. Remember that this is also the time when locations in the Northern Hemisphere experience the fewest hours of daylight. The combination of a short daylight period and less direct sunlight causes the low temperatures of the winter season.

Summer

On June 21, the Northern Hemisphere is tilted farthest toward the Sun. Therefore, sunlight strikes the Northern Hemisphere most directly, and the Sun appears at its highest altitude of the year at solar noon. It is also the time when the daylight period is longest. The combination of a long daylight period and more direct sunlight results in high temperatures during the summer season.

Spring and fall

At the equinoxes on March 21 and September 22, Earth is tilted neither toward nor away from the Sun, resulting in exactly equal periods of daylight and darkness. The sunlight striking Earth's surface is halfway between its most and least direct. The length of day and angle of insolation are between the two extremes; therefore, the temperatures are moderate in the spring and fall seasons.

CAUTION—Major Mistake Territory!

Many people think that Earth is warmest in the summer because that is when Earth is closest to the Sun. But Earth is actually closest to the Sun on January 4 during the winter season in the Northern Hemisphere and farthest from the Sun on July 3 during the summer season. Furthermore, Earth's orbit is so nearly circular that the difference in Earth's distance to the Sun only varies by about 2.5 percent from winter to summer. This is too small a change in distance to account for the great differences in winter and summer temperatures.

Temperatures are warmer in summer than in winter because (1) days are longer in the summer; therefore, Sunlight warms Earth's surface for more hours each day; and (2) Sunlight is more direct in summer; therefore, every hour of Sunlight has a greater warming effect.

Solstices and equinoxes

From December to June, the daylight period becomes longer each day and the altitude of the Sun at noon increases. On June 21, the altitude of the Sun at noon stops increasing. Therefore, this date is called the **summer solstice**, meaning summer "Sun stop." From June to December, the daylight period gets shorter each day and the altitude of the Sun at noon decreases. On December 21, the altitude of the Sun at noon stops decreasing. Therefore, this date is called the **winter solstice**, meaning winter "Sun stop." March 21 and September 22, when the daylight and darkness periods are about equal, are called, respectively, the **spring** and **fall equinox**, meaning spring and fall "equal night." This cyclic pattern of change repeats in an annual cycle.

BRAIN TICKLERS Set # 4

1. The diagram below shows the position at which the Sun appears in the sky at various times of the day to an observer on Earth.

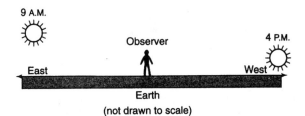

The apparent change in the position of the Sun is caused by the

a. Sun's gravity.

c. Earth's gravity.

b. Sun's rotation.

d. Earth's rotation.

2. The length of one day on Earth is determined by how long it takes

a. the Moon to revolve once.

c. Earth to rotate once.

b. the Moon to rotate once.

d. Earth to revolve once.

3. The time required for Earth to complete one revolution around the Sun is

a. one day.

c. one month.

b. one week.

d. one year.

Base your answers to questions 4–6 on the diagrams below. Diagram 1 shows Earth and the Sun's rays. Diagram 2 shows Earth in four positions, A–D, as it orbits the Sun.

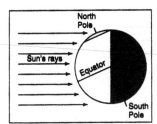

Diagram 1

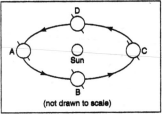

Diagram 2

4. Earth in diagram 1 corresponds to Earth in diagram 2 at position

    a. A.   **b.** B.   **c.** C.   **d.** D.

5. About how long would it take Earth to move from position A to position C in diagram 2?

    a. 12 hours   **b.** 1 day   **c.** 6 months   **d.** 1 year

6. Write the position in diagram 2 that corresponds to each of the following dates for an observer in the Northern Hemisphere:

    Winter solstice   _____   Spring equinox _____

    Summer solstice _____   Fall equinox   _____

7. In New York State there is a greater chance of precipitation falling as snow in January than in March, because in January the Northern Hemisphere is tilted

    a. toward the Sun, and temperatures are warmer.

    b. toward the Sun, and temperatures are colder.

    c. away from the Sun, and temperatures are warmer.

    d. away from the Sun, and temperatures are colder.

8. The diagram below shows Earth, the Sun, and the star constellations visible from Earth's Northern Hemisphere during different seasons of the year.

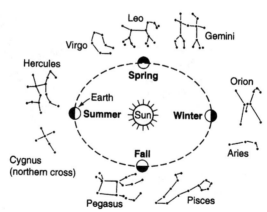

During which season is the constellation Leo not visible from Earth's Northern Hemisphere?

    a. Spring   **b.** Summer   **c.** Fall   **d.** Winter

9. The diagram below represents Earth at a specific position in its orbit. Arrows indicate radiation from the Sun. Points A–D are locations on Earth's surface.

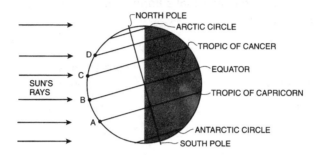

Which location would have the greatest number of daylight hours when Earth is in this position?

a. A   b. B   c. C   d. D

The map below shows the boundaries of four time zones. The clocks show the time in each of the zones at the same instant.

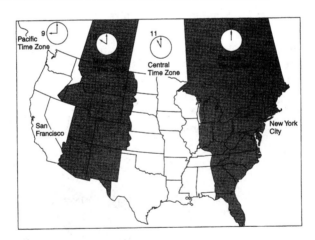

10. The time of day in New York City is always later than the time in San Francisco because Earth

a. has seasons.   c. rotates in an easterly direction.

b. orbits the Sun.   d. is tilted on its axis.

11. When it is 5 P.M. in New York City, what time is it in San Francisco?

a. noon   b. 2 P.M.   c. 8 P.M.   d. 5 A.M.

(Answers are on page 319.)

Motions of the Moon and Their Effects

As Earth revolves around the Sun, the Moon revolves around Earth every 27.32 days. The Moon's orbit is tilted at an angle of about 5° from the plane of Earth's orbit around the Sun. The Moon moves rapidly in its orbit, covering about 13° every day. As a result, every day its position against the backdrop of stars changes by 13°, or about twenty-six times its apparent diameter. The Moon also rotates on its axis once every 27.32 days. Thus, the same side of the Moon always faces Earth.

Phases of the moon

The Moon does not produce visible light of its own. The Moon is only visible by light from the Sun that is reflected from its surface. An observer on Earth is only able to see that part of the Moon that is illuminated by the Sun. As the Moon moves around Earth, different parts of the side of the Moon facing Earth are illuminated by sunlight, and the Moon passes through a cycle of phases.

Although the Moon makes one revolution in 27.32 days, it takes 29.5 days for the Moon to go through a complete cycle of phases. Why the extra two days? At the same time that the Moon is revolving around Earth, Earth is revolving around the Sun at a rate of about 1° per day. When the Moon has completed one revolution, the Sun is no longer where it was when the Moon started its revolution. In 27 days the Sun has moved 27°. Moving at about 13° per day, it takes the Moon about two days to catch up to the Sun and align with it in a new moon phase. The word *month* has its origin in "moon-th," which referred to this 29.5-day cycle of phases.

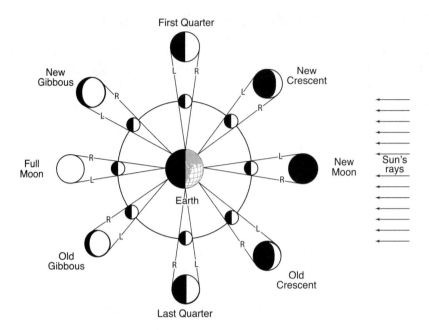

Phases of the Moon diagram.

PAINLESS TIP

The shapes of the phases of the Moon correspond to shapes of the letters in " – DOC– ."

The new moon (nothing). The new crescent and first quarter are curved on the right like a "D." The full moon is shaped like an "O." And the last quarter and old crescent are curved on the left like a "C." Then new moon (nothing).

Solar eclipses

A **solar eclipse** occurs when the Moon passes directly between Earth and the Sun, casting a shadow on Earth and blocking our view of the Sun. Both the Moon and Earth are illuminated by the Sun and cast shadows in space. Earth and its moon are very tiny compared to the Sun, so the shadows they cast are extremely long and narrow. The Moon's shadow barely reaches Earth, and the totally dark part of the shadow it casts on Earth's surface is never more than 269 kilometers in diameter. The 5° tilt of the Moon's orbit, together with the small size of its shadow, makes it easy for the shadow to miss Earth at the full and new moon phases. Thus, total eclipses of the Sun are rare.

Whether an observer sees a total or partial eclipse of the Sun depends on what part of the Moon's shadow passes over the observer. The Moon's shadow has two parts. The **umbra** is the part of the shadow in which all of the light has been blocked. In the **penumbra**, only part of the light is blocked, so the light is dimmed but not totally absent. An observer in the Moon's umbra would see a total eclipse. An observer in the penumbra would see a partial eclipse. An observer outside of the Moon's shadow would see no eclipse.

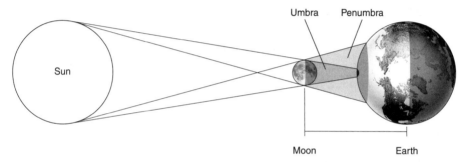

Umbra   Penumbra

Sun

Moon   Earth

Total and partial solar eclipses showing umbra and penumbra.

PAINLESS TIP

Partial = Penumbra   Partial eclipses occur when an observer passes through the penumbra.

Lunar eclipses

A **lunar eclipse** occurs when the Moon moves through Earth's shadow at full moon. If the Moon moves into Earth's umbra, a total lunar eclipse is seen. If the Moon moves into Earth's penumbra, a partial lunar eclipse is seen. When the Moon is totally in Earth's umbra, it does not completely disappear from view. While no direct sunlight reaches the Moon, some light that is bent as it passes through Earth's atmosphere reaches the Moon. Since only the long waves of red light are bent far enough to reach the Moon, the Moon glows with a dull red color during a total lunar eclipse. During a partial lunar eclipse, the Moon is only partially dimmed as Earth blocks some of the Sun's light. Partial lunar eclipses are not very impressive.

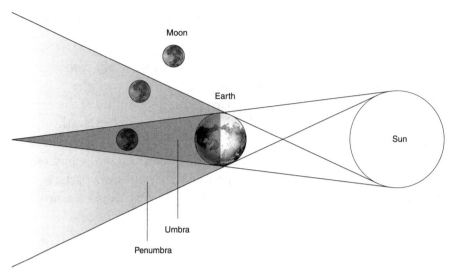

A lunar eclipse. If the Moon passes only through Earth's penumbra, a partial eclipse of the Moon would be observed. If the Moon passes through Earth's umbra, a total eclipse of the Moon is observed.

Tides

The part of Earth's surface facing the Moon is about 6,000 kilometers closer to the Moon than Earth's center. Since Earth's surface is closer to the Moon, the Moon's gravity exerts a stronger force on Earth's surface than Earth's center. Although Earth's surface is solid, it is not absolutely rigid. The Moon's gravity causes Earth's surface to flex outward forming a bulge several inches high. As the Moon moves around Earth, the bulge moves across the surface as it remains beneath the Moon.

An inches-high bulge in the bedrock spread over half Earth's surface is barely noticeable. But water is a fluid, and can flow in response to the Moon's gravity. Attracted by the Moon's gravity, water in the oceans flows into a bulge of water on the side of Earth facing the Moon. A bulge of water also forms on Earth's far side because the Moon pulls on Earth's center more strongly than on Earth's far side. This pulls Earth away from the oceans on the far side, and the water flows into this space creating a bulge. The water flows into these bulges from the area in between them creating a deep region and a shallow region in the ocean waters.

As Earth rotates on its axis, the position of the tidal bulges remains lined up with the Moon. As the rotating Earth carries a location into a tidal bulge, the water deepens and the tide rises on the beach. As the location rotates out of the tidal bulge, the water becomes shallower and the tide falls. Since there are two bulges on opposite sides of Earth, the tide rises and falls twice a day.

The Sun also produces tidal bulges in Earth's surface and oceans. At new moon and full moon, the Sun's tidal bulges and the Moon's tidal bulges align with one another and add together. The result is very high and very low tides. These are called **spring tides** because they "spring so high," not because they happen during the spring season. Spring tides occur at every new and full moon whatever the season.

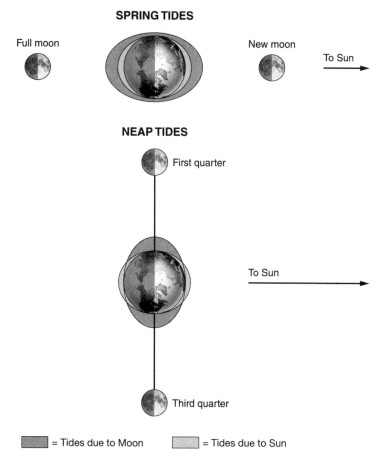

Spring tides and neap tides.

When the Sun's tidal bulges and the Moon's tidal bulges are at right angles to each other during the first and third quarter moon, they nearly cancel each other out and **neap tides** occur, in which there is very little difference between high and low tides.

BRAIN TICKLERS   Set # 5

1. Which motion causes the Moon to show phases, as seen from Earth?

   a. The rotation of the Moon on its axis

   b. The rotation of Earth on its axis

   c. The revolution of Earth around the Sun

   d. The revolution of the Moon around Earth

2. Approximately how long does it take for an observer on Earth to view a complete cycle of Moon phases?

   a. 12 hours   b. 24 hours   c. 1 month   d. 1 year

3. The diagram below shows the Moon in four positions (A–D) in its orbit around Earth.

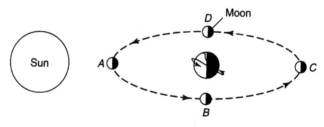

(not drawn to scale)

   A full moon could be seen by an observer when the Moon is in position

   a. A.   b. B.   c. C.   d. D.

4. The diagram below shows the relative positions of the Sun, Earth, and the Moon in space. Letters A, B, C, and D represent locations on Earth's surface.

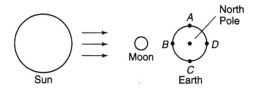

(not drawn to scale)

At which location would an observer on Earth have the best chance of seeing a total solar eclipse?

a. A   b. B   c. C   d. D

5. A total lunar eclipse will occur when the Moon moves into the

a. umbra of Earth.   c. penumbra of Earth.

b. umbra of the Moon.   d. penumbra of the Moon.

6. The diagram below shows four positions of the Moon in its orbit of Earth as seen from space above the North Pole.

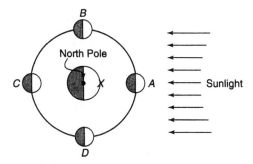

At which positions of the Moon will the highest tides occur on Earth at location X?

a. A and B   b. C and D   c. B and D   d. A and C

(Answers are on page 319.)

Wrapping up

- The earlier geocentric model of the universe has been replaced by our modern heliocentric model as a result of centuries of observation.

- Earth's rotation causes day and night and determines the time of day.

- Earth orbits the Sun in 365¼ days. The tilt of Earth's axis of rotation, combined with Earth's revolution around the Sun, produces the changes in seasons.

- Stars are huge balls of hot, glowing gases. The Sun is a star; other stars are like the Sun, but are so far away from Earth that they just look like tiny points of light.

- The solar system is made up of the Sun, the planets and dwarf planets that orbit the Sun, the satellites of those planets, and other smaller bodies that orbit the Sun, such as asteroids and comets.

- Galaxies are collections of billions of stars; nebulae are clouds of dust and gas.

- Phases of the Moon are changes in the shape of the illuminated portion of the Moon that can be seen by an observer on Earth.

- Solar eclipses occur when Earth passes through the Moon's shadow; lunar eclipses occur when the Moon passes through Earth's shadow.

- Tides are cyclic changes in sea level caused by Earth rotating through bulges in the hydrosphere due to the gravity of the Moon and Sun.

Brain Ticklers—The Answers

Set # 1, page 288

1. c   2. b   3. c   4. b   5. a

Set # 2, page 291

1. d   2. a   3. a

4. Stars are much farther from Earth than the Sun.

Set # 3, page 297

1. b   2. d   3. c   4. d   5. c   6. c   7. d

Set # 4, page 308

1. d   2. c   3. d   •   4. c   5. c

6. Winter solstice—A; spring equinox—B; summer solstice—C; fall equinox—D

7. d   8. c   9. d   10. c   11. b

Set # 5, page 316

1. d   2. c   3. c   4. b   5. a   6. d

Index